NANOCOMPOSITES AND CARBON NANOTUBES: SCIENTIFIC ANALYSIS

NANOCOMPOSITES AND CARBON NANOTUBES: SCIENTIFIC ANALYSIS

By

Maria Thompson

2016

SBS Publishers & Distributors Pvt. Ltd.

New Delhi

ISBN 13 : 9789380090634

First Published in 2016

Published by:

SBS PUBLISHERS & DISTRIBUTORS PVT. LTD.

2/9, Ground Floor, Ansari Road, Darya Ganj,

New Delhi - 110002,

INDIA

Tel: 0091.11.23289119 / 41563911

Email: mail@sbspublishers.com

www.sbspublishers.com

Preface

Nanocomposite is a multiphase solid material where one of the phases has one, two or three dimensions of less than 100 nanometers (nm), or structures having nano-scale repeat distances between the different phases that make up the material. In the broadest sense this definition can include porous media, colloids, gels and copolymers, but is more usually taken to mean the solid combination of a bulk matrix and nano-dimensional phase(s) differing in properties due to dissimilarities in structure and chemistry. The mechanical, electrical, thermal, optical, electrochemical, catalytic properties of the nanocomposite will differ markedly from that of the component materials. Carbon nanotubes (CNTs) are allotropes of carbon with a cylindrical nanostructure. Nanotubes have been constructed with length-to-diameter ratio of up to 132,000,000:1, significantly larger than for any other material. These cylindrical carbon molecules have unusual properties, which are valuable for nanotechnology, electronics, optics and other fields of materials science and technology. Carbon Nanotubes (CNT) due to their excellent mechanical, thermal and electrical properties to the best carbon fibres, have attracted composite fraternity to explore the possibility of using them as an additional reinforcement in carbon fibre reinforced polymer composites.

Editor

Contents

Chapter 1

N-TYPE THERMOELECTRIC PERFORMANCE OF FUNCTIONALIZED CARBON NANOTUBE-FILLED POLYMER COMPOSITES

Dallas D. Freeman, Kyungwho Choi, Choongho Yu

Department of Mechanical Engineering, Texas A&M University, College Station, Texas, United States of America

Corresponding Author

Email: chyu@tamu.edu

ABSTRACT

Carbon nanotubes (CNTs) were functionalized with polyethyleneimine (PEI) and made into composites with polyvinyl acetate (PVAc). CNTs were dispersed with different amounts of sodium dodecylbenzenesulfonate (SDBS) prior to the PEI functionalization. The resulting samples exhibit air-stable n-type characteristics with electrical conductivities as great as 1500 S/m and thermopowers as large as −100 μV/K. Electrical conductivity and thermopower were strongly affected by CNT dispersion, improving the properties with better dispersion with high concentrations of SDBS. This improvement is believed to be due to the increase in the number of tubes that are evenly coated with PEI in a better-dispersed sample. Increasing the amount of PEI relative to the other constituents positively affects thermopower but not conductivity. Air exposure reduces both thermopower and conductivity presumably due to oxygen doping (which makes CNTs p-type), but stable values were reached within seven days following sample fabrication

INTRODUCTION

Functional thermoelectric devices can produce electricity anywhere there is a temperature gradient. Conversely, they can be employed as refrigeration devices if current is supplied, cooling without the use of pumps or fluids. At this point, the limited efficiency of thermoelectric modules restricts their applications in both electricity generation and refrigeration to situations where longevity, space requirements, and quiet operation are of principle importance [1].

Starting in the 1990s, research into thermoelectric materials has been re-energized by new material fabrication techniques. By controlling the structure at the nano- and micro-scales, improvements have been made to the efficiency of the Bi-Te alloys, which were the state of the art for decades [2]. It is hoped that further research into nanomaterials will broaden the applicability of thermoelectric energy conversion by enhancing their performance.

The thermoelectric figure of merit, used to describe the thermoelectric effectiveness of any material, is given by:

$$Z = \frac{S^2\sigma}{\kappa}$$

Where S, σ, and κ are thermopower, electrical conductivity, and thermal conductivity, respectively [2]. Polymers typically have low thermal conductivity, which is a good starting point for a high figure of merit. However, typical polymers do not have an electrical conductivity, high enough for efficient thermoelectrics. In order to improve the electrical properties, carbon nanotubes have been used together with polymers to synthesize composites [3], [4], [5], [6], [7].

Carbon nanotubes (CNTs) have been extensively studied as potential solutions in a wide range of applications due to their unique mechanical, electrical, and geometric properties [8]. Many of the tubes, which are composed of one or more rolled sheets of the carbon honeycomb structure known as graphene, are either metallic or semiconducting, depending on the lattice vector by which they are rolled. Intrinsically n-type semiconducting nanotubes are highly susceptible to oxygen doping and become p-type in atmosphere [9], [10].

Several methods have been demonstrated for the production of air stable n-type nanotubes, including passivation of a protective film around the tubes to prevent oxygen doping, application of viologens for a direct redox reactions, and the use of metal electrodes with low work functions [11], [12], [13]. A simpler production method has also been demonstrated wherein the physical adsorption of branched polyethyleneimine (PEI) onto carbon nanotubes results in a conversion of the conducting properties from p-type back to n-type [6], [14]. Nanotubes functionalized with PEI have been used to produce p-n junctions, photovoltaic cells, and field-effect transistors [14], [15], [16], [17]. In our previous work, the thermopower of thin films composed of PEI-doped tubes was measured to be as large as -60 μV/K [6]. Such films, composed almost purely of carbon nanotubes, are not prime candidates for thermoelectric generation because their thermal energy transport may be too high due to their intrinsically high thermal conductivity [10].

CNTs were combined with polymers to synthesize composites, and their electrical and thermal properties have been evaluated in our prior work [3], [4], [5], [18]. Despite typical correlation between thermal and electrical conductivities, these composites exhibit electrical conductivities nearly as high as films composed exclusively of tubes, but still possess thermal conductivities closer to those their polymer matrices. This phenomenon results from the relative ease with which charge carriers travel across the nanotube networks by hopping. The thermal carriers, or phonons, have relative difficulty with transport because they are scattered at the CNT surfaces and at the junctions between the tubes [3]. The result is a material with high electrical conductivity and low thermal conductivity. For example, composites with p-type doped CNTs exhibited electrical conductivities as high as ~105 S/m [3]. While several studies have been conducted on polymer composites containing CNTs, air-stable n-type polymer composites with CNTs have never been reported. The overall aim of this study was to produce and test such composites and to determine the conditions which result in the best thermoelectric performance.

RESULTS AND DISCUSSION

Series 1 was composed of samples containing 20-wt% CNT, 10-wt% PEI, and a varied amount of sodium dodecylbenzenesulfonate (SDBS). As shown in the scanning electron microscope (SEM) images for three samples of different conductivities (Figure 1), the samples with higher SDBS weight percent exhibit smoother cleavages and fewer CNT pullouts (PEI wraps around CNTs, it is rather hard to see CNTs clearly in the micrographs). Both of these are characteristics of good dispersion, indicating that SDBS weight percent at a ratio of 3 to 1 with CNT is more effective at deconstructing the bundles than the smaller ratios. As the amount of SDBS goes down, the CNTs agglomerate more, forming networks around internal voids like those shown in Figure 1a. Such voids reduce overall electrical conductivity for the composite.

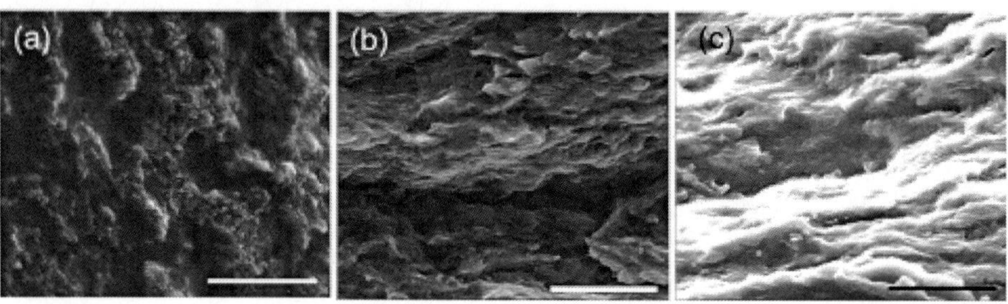

Figure 1. SEM images of cold fractured surfaces for samples with 20, 40, and 60-wt% SDBS (a, b, and c, respectively).

These samples contain 20 wt% CNTs (99% purity). Fractures become sharp and disorganized as less surfactant is included, eventually forming heterogeneous structures as seen in (a). Scale bars indicate 10 μm.

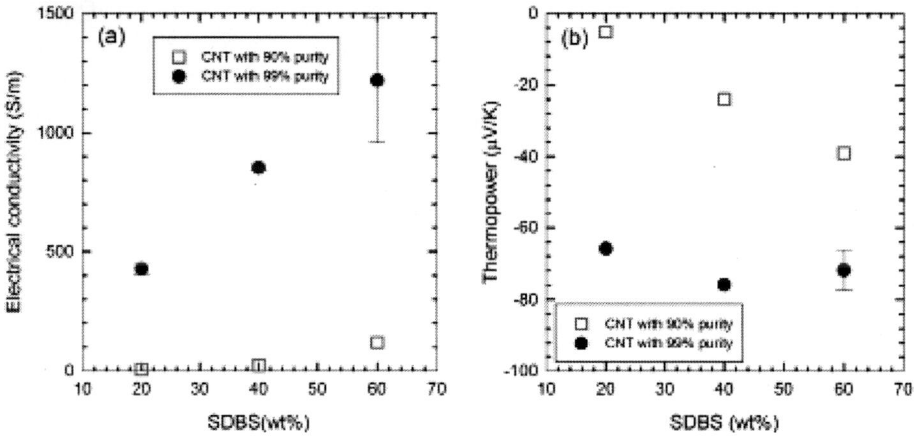

Figure 2. Electrical conductivity (a) and thermopower (b) for Series 1 samples containing 20-wt% CNT, 10-wt% PEI, and a varied amount of SDBS.

Figure 2 depicts correlation between SDBS and both thermopower and electrical conductivity. There are large differences between tubes rated as 99% and 90% purities, probably due to the lower conductivity of amorphous carbon and the catalyst impurities [7]. The heavy catalyst particles may also have a negative effect on dispersion, forcing tubes to which they are attached to fall into a precipitate. It also clearly shows strong dependence of the properties as a function of SDBS concentrations. In most materials, thermopower and conductivity have an inverse relationship, explained qualitatively by the fact that an increased amount of charge carriers will drive down the voltage potential being induced by the temperature gradient. However, our results show an increase in electrical conductivity resulted in a larger thermopower (absolute value). We believe that this can be attributed to CNT dispersion, which makes the sites for PEI doping upon better dispersion with a high concentration of SDBS.

Correlation is positive for thermopower magnitude and electrical conductivity.

The electrical conductivity of the PEI-doped samples (~103 S/m) in this study is lower than that of the samples that we previously reported [3], [4], [5]. This can be attributed to the poor electrical conductivity (an electrical insulator [19]) and the branched nature of PEI. When CNTs are embedded in the composite, PEI wraps around CNTs, hindering direct contacts between nanotubes. The tube-tube junctions play an important role in altering the electrical conductivity of such composites [5]. When electrically-conducting poly(3,4-ethylenedioxythiophene) poly(styrenesulfonate) (PEDOT:PSS) is used for dispersing CNTs, they are electrically well connected due to the presence of electrically conducting PEDOT:PSS at the junctions. Another reason is that the current study includes only 20 wt% (except one 40-wt% sample) CNTs, as opposed to ~60 wt% CNT concentration to achieve 105 S/m conductivity. Additionally, we used CNTs synthesized by a chemical vapor deposition method, which have relatively low electrical conductivity compared to other tubes synthesized by an arc-discharge or high pressure carbon monoxide (HipCo) method [4], [7].

Dispersion remains a primary consideration in the production of CNT composites to obtain desired electrical, thermal, or/and structural properties [3], [4], [5], [7], [20]. In this work, the surfactant SDBS was chosen to facilitate dispersion based on its superior performance and for its tendency to form smooth coating layers around the CNTs [21], [22]. The principal challenge specific to this research resulted from the use of PEI. This polymer contains one of the highest densities of amine groups of all polymers, and donates electrons to the nanotubes [6], [23]. The amine groups contain electron lone pairs, which are responsible for the electron donation. This PEI attachment on CNTs makes CNTs electron-rich, converting p-type CNTs into n-type. When CNTs were well dispersed, the number of nanotubes in each bundle decreases, resulting in an increase in the amount of tubes coming into physical contact with the PEI molecules, allowing for more effective doping. In other words, fewer tubes from the center of the bundles are allowed to remain p-type and reduce the magnitude of the composite thermopower value, as discussed in the later sections. Conversely, when CNTs were "not" well dispersed, a reduction in electron pathways through the material happens. Fewer pathways mean that the composite as a whole will have less conductivity than a material where the tubes were more completely dispersed.

The PEI used in this study had an average molecular weight of ~600, indicating that it was composed of molecules that were made of four to five units like the one shown in Fig. 3. Although covalent bonding has been demonstrated for the case of F-functionalized (fluorinated) CNTs, pristine CNTs like those used in this study interact with PEI dominantly via physisorption [7], [22] rather than chemisorption. In this process, the PEI molecules wrap around the CNTs and form bonds with other PEI molecules or with the opposite ends of the same molecule. It has been shown with atomic force microscope imaging that this coating of PEI molecules causes the tube to double in diameter relative to uncoated tubes [24]. In our experiments, when tubes were soaked in PEI solutions, and then filtered, rinsed, and allowed to dry, the product was heavier than the tubes prior to the process. In addition to adding mass, PEI also acts as a coagulant for CNTs in water which can counteract the effects of the surfactant. Visible tube agglomerations were observed when the ratio of PEI wt% to CNT wt% is higher than 0.5 during the PEI doping process. Such aggregation often makes the sample mechanically weak and/or hinders composite formation.

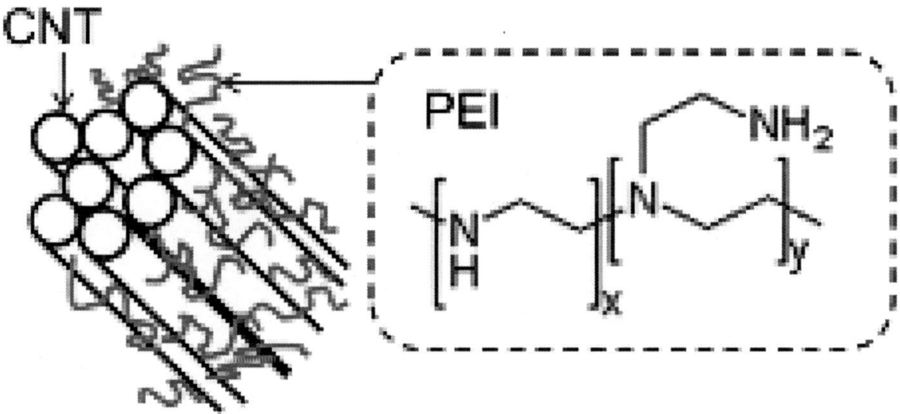

Figure 3. Schematic representation of PEI-functionalized CNT bundles in the composite and the structure of branched PEI.

The amine groups are responsible for the electron donation which converts CNTs into n-type semiconductors. When the bundles are smaller, more tubes come in direct contact with PEI, which increases the n-type thermopower values. On the other hand, composites with large bundles have fewer junctions necessary for electron transport across the composite.

Series 2 and Series 3 varied the amount of PEI, being composed of 20-wt% CNT with 20-wt% (for Series 2) or 40-wt% (for Series 3) SDBS. When the higher SDBS concentration was used, electrical conductivity was more or less improved, presumably due to the better CNT dispersion. The data are somewhat scattered but the overall values for electrical conductivity are higher for the 40-wt% SDBS samples, compared to the 20-wt% samples. The change in thermopower values is rather small due to the small absolute values as well as the weighting factor (electrical conductivity) for n- and p-type thermopower of CNTs in the composite materials, as discussed more in detail below (see equation 2). The composites contain less SDBS compared to the best properties in Series 1 (60-wt% SDBS). As we had more CNTs compared to SDBS, it is much harder to disperse CNTs, which is a part of the reasons for the scattered properties. It should be noted that, unlike p-type properties, n-type strongly depends on PEI coverages. On the other hand, the samples exhibited little change with increasing PEI beyond 10 wt%, possibly indicating saturation (Figure 4). Beyond this point, it is likely that no rooms are available for additional PEI to attach.

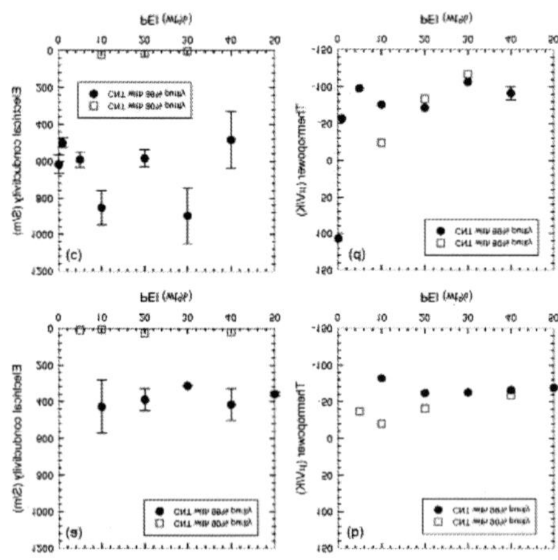

Figure 4. Effect of PEI wt% on the electrical properties for the samples containing 20-wt% (a and b: Series 2:) and 40-wt% (c and d: Series 3) SDBS.

Thermopower sign changed with PEI whose concentrations between 0.1 and 1 wt%, as shown in (d), wherein it is seen that values for thermopower do not change significantly beyond 5-wt% PEI.

The first three levels of Series 3 (0.1, 1, and 5-wt% PEI, in Figure 4(d)) demonstrate a correlation between PEI weight percent and thermopower. The sample containing only 0.1 wt% has a p-type thermopower comparable to control samples made without incorporating PEI. The 1-wt% sample exhibits n-type properties of a lower magnitude. Above 5 wt%, the samples demonstrate only a slight correlation with thermopower and additional PEI weight percent. Doping fractions by nitrogen per each carbon in CNTs were estimated. Assuming the structure of the branched PEI is shown in Figure 3 with x and y are 1 (i.e., 1 unit contains 3 nitrogen). Then, molecular weight of 5 units is ~645, which is close to the specification from the manufacturer. The n-type conversion occurred when the ratio of CNT to PEI wt% is 1:0.05 (i.e., 1 wt% PEI sample). In this case, the number of nitrogen per carbon can be calculated to be ~3×10−3. In fact, this value is within the range, 1–6×10−3 [25] suggested by an experiment with a field-effect transistor configuration made of PEI-functionalized CNT. This may explain that relatively constant thermopower when PEI is more than 10 wt%.

The effects of air on the samples (Series 4) over time are depicted in Figure 5. Both conductivity and thermopower decayed somewhat and then became more or less constant in the presence of air, owing to the increased levels of oxygen doping on the material. The simultaneous reduction in conductivity and thermopower experienced during atmospheric exposure over time can be understood in terms of a mixed carrier model, where a conductor uses both electrons and holes as charge carriers. In particular, each nanotube bundle in the composites is unlikely to have been permeated with PEI throughout the bundle. This will leave n-type tubes on the outside where the PEI is attached, while the tubes inside the bundle will still be p-type due to oxygen doping. In addition, as the sample is left exposed to the atmosphere, more oxygen molecules will infiltrate the polymer and remove more electrons from the tubes. Therefore, the entire composite is a mixture of p- and n-type charge carriers, which will recombine to have reduced carrier concentrations, resulting in a lower electrical conductivity. Assuming that both carriers are present after recombination, thermopower of a composite with mixed carriers may be described as [6]:

$$S = \frac{S_e \sigma_e + S_h \sigma_h}{\sigma} \tag{2}$$

where Se is the thermopower due to electrons (having a negative value) and Sh is the value for holes; σe and σh are the conductivities due to electrons and holes, respectively; σ is electrical conductivity due to both electrons and holes. As can be seen, an increase in hole conduction will decrease the overall thermopower. CNTs also exhibit different electrical properties depending on their chirality, or the arrangement of the graphene hexagons relative to the tube axis [8]. These electronic differences may make individual tubes more or less responsive to doping, depending on metallic or semiconducting properties.

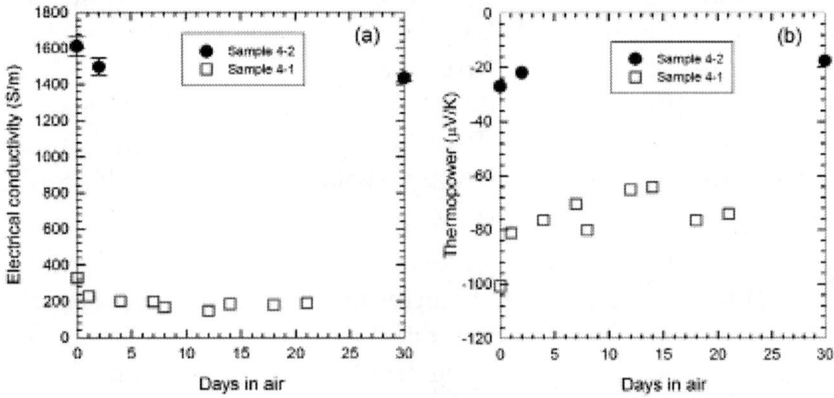

Figure 5. Change in electrical conductivity (a) and thermopower (b) as a function of time.

The elapsed time was measured after the vacuum annealing (last process for fabricating samples) for Series 4 samples.

It should be noted that in the case for pristine n-type nanotubes in vacuum, air exposure results in an increase of conductivity, as oxygen donates sufficient holes to overcome the loss in electrons [9]. In our samples shown in Figure 5, after a few days of air exposure, thermopower was saturated. It appears the PEI coating sustains n-type doping for the time periods. This suggests PEI was coated on most of CNT surfaces, deterring oxygen doping.

In the future, studies could be conducted to optimize n-type doping along with p-type doping [6], [26] by adjusting the ratios between CNT, SDBS, PEI, and other molecules or chemicals. Oxygen barriers could be used to prevent degradation of n-type properties. Eventually, a working combined p- and n-type modules can be fabricated and evaluated for the thermoelectric figure of merit [23]. This work will pave the way for lightweight, non-toxic and flexible thermoelectric cells capable of harvesting thermal energy from the human body, solar cells and a host of other areas where it is currently going to waste as well as cooling energy consuming devices.

MATERIALS AND METHODS

Carbon nanotubes, made using chemical vapor deposition by CheapTubes Inc., were used for the experiment. The manufacturer claimed the 90 wt% purity CNTs contain single wall nanotubes whose specifications are the following: outer diameter: 1–2 nm, inner diameter: 0.8–1.6 nm, length: 5–30 μm, ash: <1.5 wt%, multi-wall nanotubes: >5 wt%, amorphous carbon: <3 wt%. The 99 wt% purity CNTs contain approximately 50/50 single- and double-wall tubes whose specifications are the following: outer diameter: 1–2 nm, inner diameter: 0.8–1.6 nm, length: 3–30 μm, ash: 0 wt%, multi-wall nanotubes: <2 wt%.

For each experiment, 60 mg of CNTs were dispersed in 5~15 ml of deionized water with a prescribed amount of SDBS. Sonication was conducted in two modes. The mixture was sonicated for 30 min using pen-type sonicators (XL-2000 from Misonix and FB 120 from Fisher scientific).

Afterward, a determined amount of aqueous 5-wt% PEI (branched, M.W. 600, 99% from Alfar Aesar) solution was added with a pipette into the dispersion. It has been demonstrated that PEI attaches to nanotubes by physisorption on the tube sidewall [27]. To maximize the occurrence of physisorption and create an even coating of PEI on the nanotubes, the mixture solution was stirred for 48 hours while being maintained at a temperature of 50–60°C. The prescribed amount of PVAc (Vinnapas 401, Wacker chemicals) was subsequently added into the dispersion. The mixture was dispersed by pen-type sonicators for 30 minutes each and then poured into a 5 cm×5 cm×2 cm plastic container for casting. Films of 20–80 μm in thickness were formed as the dispersion dried, in a process that usually took between 24 and 48 hours. Dried samples were then thermally annealed in a vacuum oven at 60°C for 4 hrs in order to remove any water which had permeated the film.

Four series of samples were synthesized. In the first, the weight percent of SDBS was varied while the weight percentages of CNT and PEI were maintained at 20 and 10 wt%, respectively. The second and third series varied PEI wt% while maintaining SDBS at 20 wt% and 40 wt%, respectively, with 20-wt% CNTs. The fourth was for testing air stability with 20 or 40 wt% CNTs in the composites. All weight percentages were determined by measuring the mass of the material on a scale before including it in the sample. The weight percent of Vinnapas was varied in each series to make up whatever difference was left between the total of the other three weight percentages and 100 percent. A summary of the experimental details is included in Table 1.

Table 1. List of samples with the concentrations of CNT, SDBS, PEI, PVAc as well as the purity of CNT.

Series	sample Number	CNT wt%	SDBS Wt%	PEI Wt%	PVAc Wt%	CNT Purity%
1	1	20	20	10	50	90
	2	20	40	10	30	90
	3	20	60	10	10	90
	4	20	20	10	50	99
	5	20	40	10	30	99
	6	20	60	10	10	99
2	1	20	20	5	55	90
	2	20	20	10	50	90
	3	20	20	20	40	90
	4	20	20	40	20	90
	5	20	20	10	50	99
	6	20	20	20	40	99
	7	20	20	30	30	99

	8	20	20	40	20	99
	9	20	20	50	10	99
3	1	20	40	10	30	90
	2	20	40	20	20	90
	3	20	40	30	10	90
	4	20	40	0.1	39.9	99
	5	20	40	1	39	99
	6	20	40	5	35	99
	7	20	40	10	30	99
	8	20	40	20	20	99
	9	20	40	30	10	99
	10	20	40	40	0	99
4	1	20	40	40	0	99
	2	40	40	10	10	90

After fabrication, each sample was tested for conductivity and thermopower values. To this end, a rectangular test sample of the dried and annealed film was removed from the plastic container. Conductive silver paint was applied to the sample strip to minimize electrical contact resistance and the relevant dimensions were measured using a micrometer. Details can be found in our previous publications [3], [4], [5], [6].

AUTHOR CONTRIBUTIONS

Conceived and designed the experiments: CY DF. Performed the experiments: DF KC CY. Analyzed the data: CY DF. Contributed reagents/materials/analysis tools: DF KC CY. Wrote the paper: DF CY KC.

REFERENCES

1. DiSalvo F (1999) Thermoelectric cooling and power generation. Science 285: 703–706.
2. Majumdar A (2004) Thermoelectricity in semiconductor nanostructures. Science 303: 777–778. doi: 10.1126/science.1093164
3. Yu C, Choi K, Yin L, Grunlan JC (2011) Light-weight flexible carbon nanotube based organic composites with large thermoelectric power factors. ACS

Nano 5: 7885–7892. doi: 10.1021/nn202868a

4. Yu C, Kim YS, Kim D, Grunlan JC (2008) Thermoelectric behavior of segregated-network polymer nanocomposites. Nano Lett 8: 4428–4432. doi: 10.1021/nl802345.
5. Kim D, Kim Y, Choi K, Grunlan JC, Yu C (2010) Improved thermoelectric behavior of nanotube-filled polymer composites with poly(3,4-ethylenedioxythiophene) poly(styrenesulfonate). ACS Nano 4: 513–523. doi: 10.1021/nn9013577
6. Ryu Y, Freeman D, Yu C (2011) High electrical conductivity and n-type thermopower from double-/single-wall carbon nanotubes by manipulating charge interactions between nanotubes and organic/inorganic nanomaterials. Carbon 49: 4745–4751. doi: 10.1016/j.carbon.2011.06.082
7. Ryu Y, Yu C (2012) Dramatic electrical conductivity improvement of carbon nanotube networks by de-bundling and hole-doping with chlorosulfonic acid. J Mater Chem 22: 6959–6964. doi: 10.1039/c2jm16000e
8. Meyyappan M (2005) Carbon nanotubes: Science and applications. Florida: CRC Press LLC.
9. Collines PG, Bradley K, Ishigami M, Zettl A (2000) Extreme oxygen sensitivity of electronic properties of carbon nanotubes. Science 287: 1801. doi: 10.1039/c2jm16000e
10. Yu C, Shi L, Yao Z, Li D, Majumdar A (2005) Thermal conductance and thermopower of an individual single-wall carbon nanotube. Nano Lett 5: 1842–1846. doi: 10.1021/nl051044e
11. Kaminishi D, Ozaki H, Ohno Y, Maehashi K, Inoue K (2005) Air-stable n-type carbon nanotube field-effect transistors with si3n4 passivation films fabricated by catalytic chemical vapor deposition. Appl Phys Lett 86. doi: 10.1039/c2jm16000e
12. Kim SM, Jang JH, Kim KK, Park HK, Bae JJ, et al. (2008) Reduction-controlled viologen in bisolvent as an environmentally stable n-type dopant for carbon nanotubes. J Am Chem Soc 131: 327–331. doi: 10.1021/ja807480g
13. Zhang Z, Liang X, Wang S, Yao K, Hu Y, et al. (2007) Doping-free fabrication of carbon nanotube based ballistic cmos devices and circuits. Nano Lett 7: 3603–3607. doi: 10.1021/nl0717107
14. Shim M, Javey A, Kam NWS, Dai H (2001) Polymer functionalization for air-stable n-type carbon nanotube field-effect transistors. J Am Chem Soc 11512–11513. doi: 10.1021/ja0169670
15. Yamaguchi I, Yamamoto T (2004) Soluble self-doped single-walled carbon nanotube. Mater Lett 58: 598–603. doi: 10.1021/nl0717107
16. Li Z, Saini V, Dervishi E, Kunets VP, Zhang J, et al. (2010) Polymer functionalized n-type single wall carbon nanotube photovoltaic devices. Appl Phys Lett 96: 033110:1–3. doi: 10.1021/nl0717107
17. Lin Y-M, Appenzeller J, Knoch J, Avouris P (2005) High-performance carbon nanotube field-effect transistor with tunable polarities. IEEE T Nanotechnol 4: 481–489. doi: 10.1109/tnano.2005.851427

18. Park W, Choi K, Lafdi K, Yu C (2012) Influence of nanomaterials in polymer composites on thermal conductivity. J Heat Transf-T ASME 134: 041302:1-7. doi: 10.1109/tnano.2005.851427

19. Herlem G, Reybier K, Trokourey A, Fahys B (2000) Electrochemical oxidation of ethylenediamine: New way to make polyethyleneimine-like coatings on metallic or semiconducting materials. J Electrochem Soc 147: 597–601. doi: 10.1109/tnano.2005.851427

20. Moore VC, Strano MS, Haroz EH, Hauge RH, Smalley R (2003) Individually suspended single walled carbon nanotubes in various surfactants. Nano Lett 3: 1379–1382. doi: 10.1109/tnano.2005.851427

21. Islam MF, Rojas E, Bergey DM, Johnson AT, Yodh AG (2003) High weight fraction surfactant solubilization of single-wall carbon nanotubes in water. Nano Lett 3: 269–273. doi: 10.1109/tnano.2005.851427

22. Dillon EP, Crouse CA, Barron AR (2008) Synthesis, characterization, and carbon dioxide adsorption of covalently attached polyethyleneimine-functionalized single-wall carbon nanotubes. ACS Nano 2: 156–164. doi: 10.1021/nn7002713

23. Yu C, Murali A, Choi K, Ryu Y (2012) Air-stable fabric thermoelectric modules made of n- and p-type carbon nanotubes. Energy Environ Sci In print, DOI: 10.1039/C2EE22838F.

24. Liao K-S, Wang J, Früchtl D, Alley NJ, Andreoli E, et al. (2010) Optical limiting study of double wall carbon nanotube–fullerene hybrids. Chem Phys Lett 489: 207–211. doi: 10.1039/c2ee22838f

25. Moonsub Shim AJ, Nadine Wong SK, Hongjie D (2001) Polymer functionalization for air-stable n-type carbon nanotube field-effect transistors. J Am Chem Soc 11512–11513. doi: 10.1021/ja0169670

26. Ryu Y, Yu C (2011) The influence of incorporating organic molecules or inorganic nanoparticles on the optical and electrical properties of carbon nanotube films. Solid State Comm 151: 1932–1935. doi: 10.1039/c2ee22838f

27. Mamedov AA, Kotov NA, Prato M, Guldi DM, Wicksted JP, et al. (2002) Molecular design of strong single-wall carbon nanotube/polyelectrolyte multilayer composites. Nat Mater 1: 190–194. doi: 10.1038/nmat747

Chapter 2

HIGHLY DOPED CARBON NANOTUBES WITH GOLD NANOPARTICLES AND THEIR INFLUENCE ON ELECTRICAL CONDUCTIVITY AND THERMOPOWER OF NANOCOMPOSITES

Kyungwho Choiand Choongho Yu

Department of Mechanical Engineering, Texas A&M University, College Station, Texas, United States of America

Corresponding Author

Email: chyu@tamu.edu

INTRODUCTION

Carbon nanotubes (CNTs) have been considered as promising candidates for various applications including field effect transistors (FETs) [1], [2], touch screens [3], [4], field emission displays (FEDs) [5], [6], and solar cells [7]–[9] due to their outstanding electrical properties. Recently, CNTs were used as fillers in polymer composites and their electrical conductivities were orders of magnitude higher than other polymer composites with conductive fillers [10]–[14]. It has been shown that the electrical conductivity can be dramatically increased as a function of nanotube loadings in the composites. The highest electrical conductivity was obtained with 60 wt%, but the conductivity was decreased with composites containing CNTs more than 60 wt% [14]. The reduction in electrical conductivity is due to CNT aggregations caused by the insufficient amount of dispersants (which cannot be increased due to high CNT loadings). The optimum ratio of CNT to stabilizer for high electrical conductivity was found to be 3:2. This means that the maximum CNT concentration should not be larger than 60 wt% for improving conductivity.

In this work, we demonstrate that nanoparticles can be incorporated on nanotube surfaces in order to further improve the electrical conductivity of nanocomposites. This also provides the influence of spherical-shape metal nanoparticles on the electrical conductivity of polymer composites. Nanoparticles can be precipitated on nanotubes by galvanic displacement or reduction potential differences between nanotubes and nanoparticles [15]. When nanoparticles are precipitated on nanotubes, charge transfer between them occurs, altering electrical transport properties of the nanotubes. Such property changes are similar to semiconductor doping with an acceptor impurity. For convenience, we shall therefore refer to the nanoparticle precipitation process as 'doping'. In this paper, we particularly studied the influence of gold nanoparticle incorporation into CNT-filled composites on their electrical properties.

The electrical properties were measured with three different gold concentrations, 10, 15, or 20 vol%, in order to identify the effect of p-type doping on the conductivity, dispersion, and microstructure of the resulting composites. The maximum electrical conductivity was measured to be ~6×105 S/m with 60 vol% of single wall carbon nanotubes (SWCNTs) and 15 vol% of gold nanoparticles. This electrical conductivity is orders of magnitude higher than those of other polymer composites with comparable concentration of gold nanoparticle (10−4~102 S/m) [16]–[18]. With the variable range hopping model, the effect of p-type doping was also analyzed. Furthermore, composites containing different type CNTs were also synthesized, and their electrical properties and microstructures were presented in the following sections.

RESULTS AND DISCUSSION

All samples contain 60-vol% CNTs and 10-vol% PVAc, and the rest 30 vol% was PEDOT:PSS or PEDOT:PSS with gold (2:1, 1:1, and 1:2 ratios), as listed in Table 1. For the samples containing SWCNT (Sample #: 1~6), many nanotubes in the sample with 20-vol% PEDOT:PSS were embedded (Figure 1A) whereas the samples with 15- and 10-vol% PEDOT:PSS show more nanotubes separated from the polymer (Figure 1B and 1C), presumably due to less stabilizers. Gold nanoparticles were observed in the sample with 20-vol% gold (Figure 1C). Two PVAc polymer with different Tg (Vinnapas 401 and 600BP) were used, but we did not find any noticeable differences in microstructures. The films made from Vinnapas 600BP were more flexible than those made from Vinnapas 401 at room temperature due to the higher Tg of Vinnapas 401 than that of 600BP.

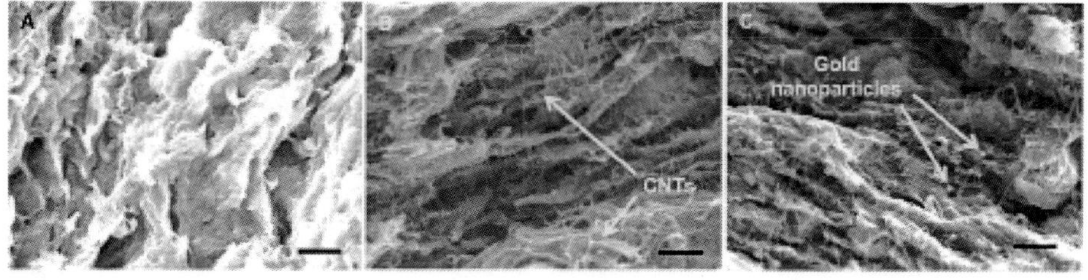

Figure 1. Cold-fractured cross sections of Sample 4 (A), Sample 5 (B), Sample 6 (C) (see

Table 1). With increasing gold vol% and decreasing PEDOT:PSS vol%, more CNTs were pulled out from the surface. The arrows indicate CNTs and gold nanoparticles. All scale bars indicate 1 μm.

Table 1. List of the composites with all contents and their vol%. doi:10.1371/journal.pone.0044977.t001

Sample number	CNT type	CNT vol%	PEDOT:PSS vol%	Au vol%	PVAc vol%		Drying time (hr) at 80°C
					401	600BP	
1	HSWCNT	60	20	10	10	-	2

2	HSWCNT	60	15	15	10		2
3	HSWCNT	60	10	20	10		2
4	HSWCNT	60	20	10		10	6
5	HSWCNT	60	15	15		10	6
6	HSWCNT	60	10	20		10	6
7	MWCNT	60	30	-	-	10	6
8	CSWCNT	60	30			10	6
9	MWCNT	60	15	15		10	6
10	CSWCNT	60	15	15		10	6

Figure 2A shows the electrical properties of Sample 1~6. The electrical conductivity was increased when the gold content was increased from 10 (Sample 1 and 4) to 15 vol% (Sample 2 and 5). The highest electrical conductivity was measured to be ~6×105 S/m with 15-vol% PEDOT:PSS, 15-vol% gold, and 60-vol% SWCNT. This value is orders of magnitude higher than those of other nanotube-filled polymer composites [10], [11] and shows ~500% improvement compared to our previous work with similar amounts of SWCNT and PH1000 (~9×104 S/m) [14]. It is likely that the electrical conductivity of gold is not the only reason that we obtained such high electrical conductivity from the composites. This is because the typical percolation threshold of gold nanoparticles is ~30 vol% in polymer composites [18], which is larger than the maximum gold concentration (20 vol%) in our experiments. In other words, when the concentration of the nanoparticles is lower than the percolation threshold, the mean distance between the nanoparticles is too large to have connected gold networks. For example, Devasdoss et al. showed that the maximum electrical conductivity is 8×10−8 S/m with a composite containing gold nanoparticles (mole ratio of 4.95×10−2) and metallopolymer [16]. Podhaecka et al. reported that the electrical conductivity of a composite with gold nanoparticles (~10 vol%) and poly(3-octylthiophene) is 10−4 S/m [17]. A high gold nanoparticle concentration, 40 vol% well above the percolation threshold in poly-4-vinyl pyridine matrices resulted in only ~102 S/m [18]. Such lower electrical conductivities suggest the high electrical conductivity from our samples is likely from p-type doping on nanotubes by the nanoparticles.

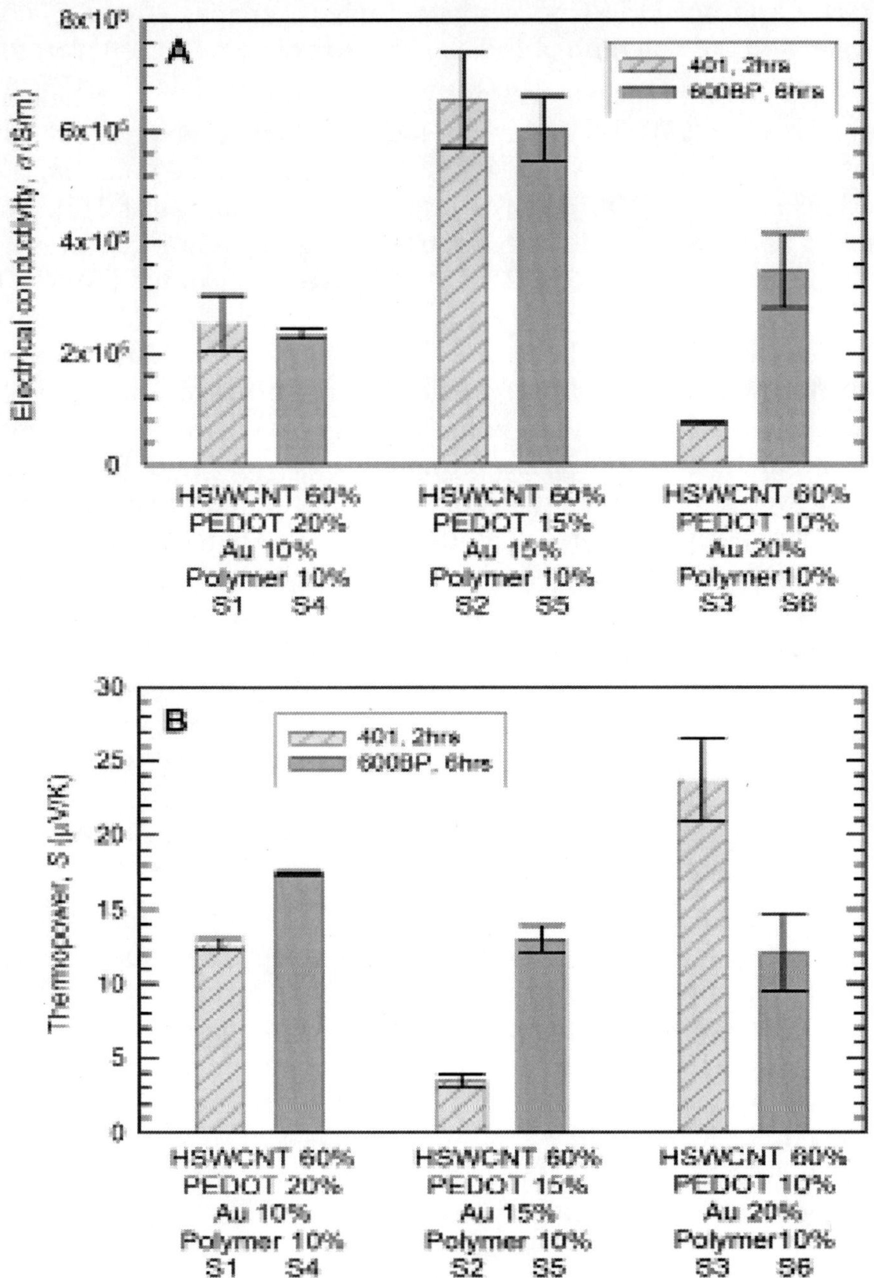

Figure 2. Electrical conductivity (A) and thermopower (B) of Sample 1~6 (see Table 1).

The vol% of HSWCNT and polymer (Vinnapas 401 or BP600) was 60 and 10. The ratio of PEDOT:PSS to gold was 2:1, 1:1, or 1:2.

Gold nanoparticles are easily precipitated by spontaneous reduction [9], [15], [19] due to the larger reduction potential of gold ions ([AuCl4]−+3e−→Au(s)+4Cl−, standard electrode potential (E0) = +0.93~1.002 V) [15], [20]–[22] than those of nanotubes [15]. This causes nanotubes to donate electrons to gold, thereby increasing hole carrier concentrations [23], [24]. The work functions of SWCNTs (4.5~5.0 eV) [25], [26] and MWCNTs (4.3~4.95 eV) [25], [27] are also smaller than that of gold (5.1~5.47 eV [28], [29]), making electrons transferred from nanotubes to gold nanoparticles. The electrical conductivity of the sample with 20-vol% gold (sample 3 and 6) is lower than those of the samples containing 15-vol% gold. The inferior conductivity with the higher gold concentration also suggests that gold nanoparticles themselves did not make percolated paths or significantly affect the electrical conductivity of our composites. We believe this is due to the poor nanotube dispersion caused by the small volume fraction of PEDOT:PSS, which de-bundles and disperses carbon nanotubes in water. When large carbon nanotube bundles are present in the composites, the number of tube-tube junctions decreases, resulting in electrically more resistive nanotube networks [14], [30]. Furthermore, the composite contains more pores because large bundles are not readily embedded in the polymer due to increased stiffness.

Two different annealing conditions (2 hrs and 6 hrs at 80°C) were tested to identify any changes in electrical properties. The longer annealing time made the sample mechanically stronger but the electrical conductivities of the samples containing 10- or 15-vol% gold are not strongly dependent on the drying condition. When the gold concentration was increased to 20 vol% (S3 and S6), the longer drying time resulted in a higher electrical conductivity. Sample 3 was particularly weaker than Sample 6, which may have affected the electrical conductivity. It should be noted that the PVAc did not alter the electrical properties significantly. Two different composites containing 60-wt% SWCNT and 30-wt% PH1000 with 10-wt% PVAc showed similar conductivities, ~9×104 S/m for Vinnapas 401 and ~8.4×104 S/m for Vinnapas BP600.

Figure 2B depicts thermopower values of Sample 1~6, which were inversely proportional to the electrical conductivities. These values are lower than those of the samples containing 60 wt% SWCNT (30~40 μV/K) [14], but higher than that of gold (1.94 μV/K at room temperature) [31]. This is another evidence that gold nanoparticles were not percolated. Sample 2 has the smallest thermopower value, which may be due to the highest electrical conductivity and shorter annealing time (mechanically weaker than Sample 5). We believe that the smaller thermopower than those of similar composites without gold can be attributed to doping.

Here, we analyzed that the influence of the gold doping on the electrical conductivity of the nanotube networks. The electrical conductivity of a composite with a high nanotube loading can be analyzed with a parallel resistance model [14], [19] and the variable range hopping model [32], [33].

The parallel resistance model describes the electrical conductivity (σc) of a composite:

$$\sigma_c = \phi_{CNT}\sigma_{CNT} + \phi_{PEDOT}\sigma_{PEDOT} + \phi_{polymer}\sigma_{polymer}$$

where σ_{CNT}, σ_{PEDOT}, and $\sigma_{polymer}$ are the electrical conductivity of nanotube networks in the composite, PEDOT:PSS, and PVAc, respectively. Also, Φ denotes the volume fraction of each material. Here, $\sigma_{polymer}\approx 0$ because the PVAc polymer is electrically insulating (less than 100 S/m) whereas the value of σ_{PEDOT} was directly measured with 100% of PEDOT:PSS film (~10^2 S/m, without dimethyl sulfoxide (DMSO) doping). Note that the electrical conductivity of PEDOT:PSS film doped with 5 wt% of DMSO was reported as ~104 S/m [34]–[36]. In our experiments, PEDOT:PSS was not doped with DMSO in order not to reduce thermopower of PEDOT:PSS. The nanotubes in our composites can be assumed to be three dimensional (3D) networks and the electrical conductivity of nanotube mat, σ_{CNT}, can be described by the 3D variable range hopping model [32], [33].

$$\sigma_{CNT}(T) = \sigma_o \exp\left[-\left(\frac{T_1}{T}\right)^{1/1+d}\right]$$

σ_o is a constant, which represents the saturated electrical conductivity of nanotube networks when the temperature effect on electron carriers is negligible at infinite temperature. T_1 is related to the energy barrier for electron hopping through tube-tube junctions. T is temperature. When d = 3, it represents bulk conduction of pure carbon nanotube mats [33]. The major difference between σ_{CNT} and σ_o comes from tube-tube junctions. σ_{CNT} is for CNT networks with polymers between nanotube junctions whereas σ_o is for pure tube-tube junctions without any materials in between (intrinsic properties without considering the junction effects). Therefore, it is possible to obtain the influence of the p-type doping on the electrical conductivity of the nanotube networks by comparing σ_o with (indicated by Au subscript) and without (indicated by NoAu subscript) gold nanoparticles, as shown in the following equations.

$$\sigma_{o,NoAu} = \sigma_{CNT,NoAu} \exp\left[-\left(\frac{T_1}{T}\right)^{\frac{1}{4}}\right]^{-1},$$

$$\sigma_{o,Au} = \sigma_{CNT,Au} \exp\left[-\left(\frac{T_1}{T}\right)^{\frac{1}{4}}\right]^{-1}$$

The normalized factors can be obtained from $\sigma_{o',Au}/\sigma_{o,NoAu'}$ and it is possible to estimate the influence of the gold doping on electrical conductivity. From Eq. (3), the normalized factor is:

$$\frac{\sigma_{o,Au}}{\sigma_{o,NoAu}} = \frac{\sigma_{CNT,Au}}{\sigma_{CNT,NoAu}}$$

The normalized factor is independent of T1 or d. Here, the electrical conductivity of nanotube networks, $\sigma_{o,NoAu}$ was referred from the electrical conductivity of HSWCNT mats (~2.5×105 S/m at room temperature, highest conductivity from SWCNT mats, to our best knowledge) [37].

The composites with similar compositions in our previous work [14] were analyzed to obtain T1 values as a function of nanotube concentration at 300 K. Here, PEDOT:PSS were also used to debundle and stabilize CNTs in water, making tube-tube junctions similar to those of the composites in this study. The composites contain 35~75 wt% SWCNTs with PEDOT:PSS and PVAc. The carbon nanotube wt% was converted into vol% and plotted in Figure 3A (hollow square) because the density of gold used in this work is one order higher than the polymers and CNTs. The conversion enables us to properly compare properties of the composites containing the same CNT concentrations, as described below. Then, the composite electrical conductivity (σc) in Eq. (1), as shown in Figure 3A (filled circles), was used with σ_{CNT} in Eq. (2) to find T1. Here, T1 at 60 vol% SWCNT concentration was obtained to be 6.06 K from the linear interpolation of 54.6 vol% (60 wt%) and 65.8 vol% (70 wt%). When we assume the tube–tube junctions are similar after the gold nanoparticle incorporation, T1 values can be used for our composites in this study. Then, we can estimate the electrical conductivity of the gold-decorated nanotubes by using Eq. (1) and (2).

In other words:

$$\sigma_{o,Au} = \frac{(\sigma_c - \sigma_{PEDOT}\phi_{PEDOT})}{\phi_{CNT}} \exp\left[-\left(\frac{T_1}{T}\right)^{\frac{1}{4}}\right]^{-1}$$

Figure 3B depicts the normalized factor (left y axis), which describes the electrical conductivity normalized by the values without gold doping (σ_o,A_u/σ_o,N_{oAu}). The electrical conductivity of the nanotube network with 15-vol% gold nanoparticles was increased by a factor of ~4. However, the electrical conductivity of the composite with 20 vol% of gold nanoparticles (Sample 3) was decreased, presumably due to poor nanotube dispersions caused by a lack of the dispersant (PEDOT:PSS) and the high concentration of gold nanoparticles.

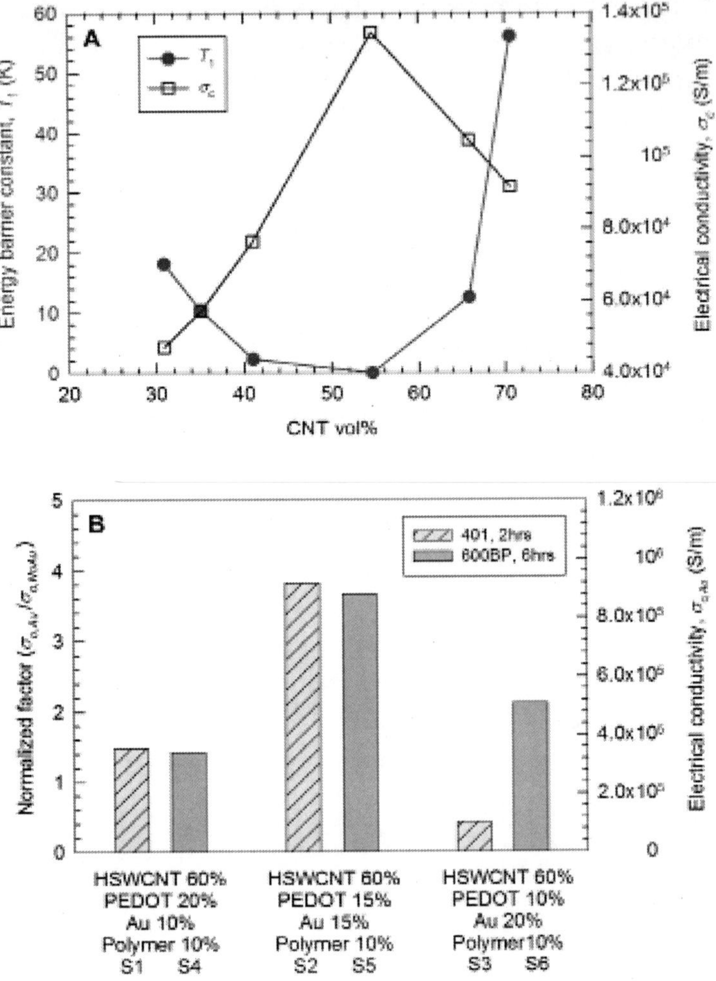

Figure 3. (A) The energy barrier constant, T1 in Eq. (2) as a function of CNT vol% (red filled circles).

(B) The normalized factor (σo,Au/σo,NoAu) that indicates the effect of gold doping for CNT networks and electrical conductivity of gold-incorporated CNT networks (σo,Au) for Sample 1~6. T1 was found from the electrical conductivities (σc) of similar composites (blue hollow squares) containing 60-wt% HSWCNT, 30-wt% PEDOT:PSS, 10-wt% PVAc from our previous work [14]. For σo,Au and σo,NoAu, tube-tube junction resistances in CNT networks were not considered.

We also used different type nanotubes (MWCNT or CSWCNT) to identify the influence of the nanotubes on the electrical properties. Samples with 60-vol% CNT, 30-vol% PEDOT:PSS, and 10-vol% polymer emulsion (Vinnapas 600BP) were prepared without gold (Sample 7 and 8). With 15-vol% gold, PEDOT:PSS was reduced to 15 vol% (Sample 9 and 10). We found that MWCNT/CSWCNT-composites containing 15-vol% gold have higher electrical conductivities, compared to the composites containing 10 and 20-vol% gold. Sample 7 and 8 (without gold) show relatively smooth and uniform cross sections, as shown in the scanning electron micrographs of Figure 4A and 4B. More nanotubes were pulled out from the polymer with CSWCNTs (Figure 4B) than MWCNTs. This may be from inferior dispersions (i.e., more aggregations) of MWCNTs compared to SWCNTs as well as from shorter lengths of MWCNTs (1~12 μm) [38] than SWNTs (5~30 μm) [39]. Additionally, the number of MWCNTs is less than that of SWCNTs due to the higher density of MWCNTs. With 15-vol% gold, relatively large gold particles were observed (Figure 4C and 4D). From the energy dispersive X-ray Spectroscopy analysis, it was confirmed that the particles are comprised of gold.

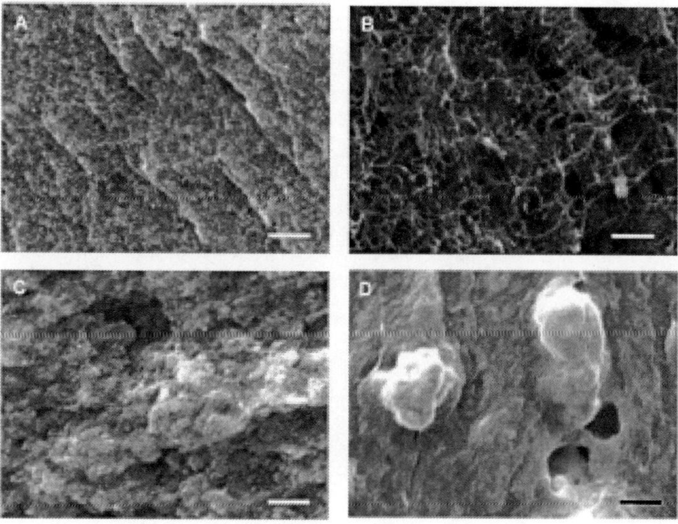

Figure 4. Cold-fractured cross sections of Sample 7 (A), Sample 8 (B), Sample 9 (C), and Sample 10 (D) (see Table 1).

Long CSWCNTs were pulled out from the polymer whereas short MWCNTs were aggregated together in (A) and (B). Relatively rough surfaces with aggregated gold particles were observed in (C) and (D). All scale bars indicate 1 μm.

The electrical conductivities of the composites containing three different types of nanotubes were compared in Figure 5A. The composite with MWCNT shows ~4×103 S/m, which is inferior to that of the HSWCNT sample (~2×104 S/m). In CNT networks, the electrical conductivity is often governed by tube-tube junctions [30]. When MWCNTs are used, the number of the junctions is small compared to those of SWCNT networks, diminishing electrons transport across the junctions. The large diameter of MWCNTs causes much smaller surface areas than SWCNTs. Moreover, the aggregation of MWCNTs can also be attributed to the inferior conductivity of the MWCNT sample. After replacing 15-vol% PEDOT:PSS with 15-vol% gold in these composites, the electrical conductivities were dramatically increased to ~7×104 and ~9×104 S/m for the MWCNT/gold and CSWCNT/gold samples, respectively. Nevertheless, these values are still lower than that of the HSWCNT/gold sample. It has been reported that the intrinsic electrical conductivity of HSWCNT is higher than that of CSWCNT (approximately one order difference) [40], generally due to the higher concentration of metallic nanotubes in HSWCNT [40]. In addition, the presence of more defects such as carbonaceous particles on the surface of the CSWCNT compared to HSWCNT may cause an increase in the contact resistance between nanotubes [41]. The large difference in the electrical conductivities of the composites with CSWCNT and HSWCNT also shows that CNT networks are the electron paths rather than gold nanoparticles. Note that the electrical conductivity of bulk gold (~4×107 S/m at 300 K) [42] is at least two-order higher than that of our composites containing CSWCNT and 15-vol% gold. The thermopower values of the composites with gold nanoparticles were measured to be less than a half of those without gold, due to the large improvement in electrical conductivity by p-doping of CNT (Figure 5B).

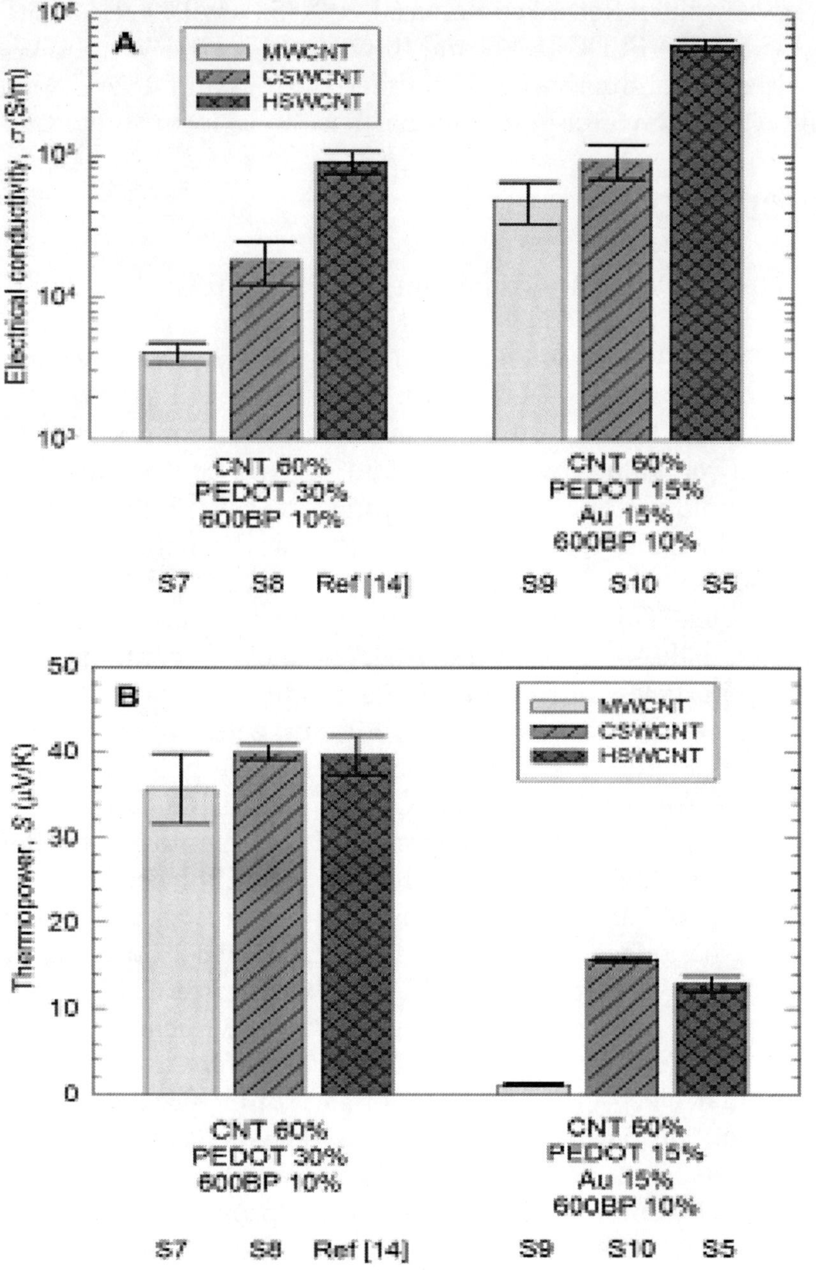

Figure 5. Electrical conductivity (A) and thermopower (B) of Sample 7~10 along with those of Sample 5 and a sample in Ref. 14 for comparison.

The sample in Ref. 14 contains 60-wt% (54.6 vol%) HSWCNT, 30-wt% (35.5 vol%) PEDOT:PSS, and 10-wt% (9.9 vol%) PVAc. HSWCNT show higher conductivities than the MWCNT- and CSWCNT-filled composites, even after gold nanoparticles were incorporated.

CONCLUSION

Polymer composites containing SWCNTs grown by a HipCo or CVD process or MWCNTs with PEDOT:PSS and PVAc. The vol% of CNT and PVAc was 60 and 10, respectively. The concentration of gold nanoparticles was 0, 10, 15, and 20 vol%, and the rest was occupied by PEDOT:PSS. Their electrical conductivities and thermopower values were measured for the composites without and with gold nanoparticles for doping CNTs. With the doping, the electrical conductivity of the composites was dramatically increased to ~6×105 S/m by replacing 15-vol% PEDOT:PSS with gold nanoparticles. This electrical conductivity is orders of magnitude higher than those of other polymer composites containing CNTs and gold particles. Furthermore, the conductivity is ~500% higher than those of similar composites without gold nanoparticles. We believe this is due to p-type doping caused by gold nanoparticles when they are precipitated on CNTs. A variable range hopping model with a parallel resistance model was employed to identify the change in the electrical conductivity of CNT networks in the composites. We also observed that the composites containing 20-vol% gold nanoparticles decreased the electrical conductivity due to the inferior CNT dispersions. This result indicates CNT dispersion with a proper amount of CNT dispersants is crucial to maximize electrical conductivity. Additionally, three different CNTs resulted in dissimilar electrical properties for the composites, showing that the intrinsic properties of the CNTs and dispersion are important factors. This study demonstrates nanoparticles can be used for doping CNTs to manipulate the electrical properties of CNT-filled polymer composites.

MATERIALS AND METHODS

We used three different-type CNTs: SWCNTs synthesized by a high pressure carbon monoxide (HipCo) process (HSWCNT) [43] and a chemical vapor deposition (CVD) method (CSWCNT) [39] as well as multi-wall carbon nanotubes (MWCNT) by a CVD method [38]. CNTs were added to deionized water (~20 ml), and the solution was sonicated with an ultrasonic homogenizer (Microson XL2000, Misonix, Inc.) for 30 minutes with 50 W power. A gold ion solution was separately prepared by adding chloroauric acid (HAuCl4, Alfa Aesar, 99.9%) to deionized water (1~2 ml), and then poured into the CNT solution, followed by 30 min sonication. Subsequently, an aqueous poly (3, 4-ethylenedioxythiophene) poly (styrenesulfonate) (PEDOT:PSS, Clevios PH1000, H. C. Starck) solution was added to the mixture, followed by 15 min sonication. PEDOT:PSS plays a role in de-bundling and dispersing CNTs in water. Finally, poly (vinyl acetate) (PVAc) emulsions were added to the mixture, followed by another 15 min sonication. Two different PVAc emulsions, Vinnapas 401 and 600BP (Wacker chemical, Co.) were used. They have different glass transition temperatures (Tg): −15 and −40°C for Vinnapas 401 and 600BP, respectively. The polymer particles in the emulsion vary in size from 0.14~3.5 μm in diameter with an average diameter of ~650 nm. The total weight including water is typically 25 g. The aqueous mixture was then poured into a 26 cm2 plastic container and dried for 48 hrs under an ambient condition in a fume hood. During the drying process, the plastic container was placed on a rotating turntable (3 rpm). Sidewalls were made on the turntable in order to avoid non-uniformity of the solid contents due to air flow in the fume hood. The solid composite was then baked in a vacuum oven at 80°C for 2 or 6 hrs. The baking process helps making strong binding between nanotubes and polymers as well as removing micro voids in the composite. Finally, fully dried composites were placed in a vacuum desiccator for 24 hs in order to completely remove residual water from the composite. The thickness of the composite ranged from 27 to 40 μm.

Table 1 shows a list of samples and vol% of the materials in the composite. The actual weights of the materials are the following. For the samples containing SWCNTs with 10-, 15-, and 20-vol% gold, the weights of HAuCl4 respectively were 0.1094 g, 0.1263 g, and 0.1368 g; the weights of CNT respectively were 0.0256 g, 0.0197 g, and 0.0160 g; the weights of PH1000 respectively were 0.4648 g, 0.2682 g, and 0.1453 g; the weights of PVAc respectively were 0.0071 g, 0.0055 g, and 0.0044 g. The solid contents of the aqueous PH1000

[44] and PVAc [12] solutions respectively were 1.5 and 55.16 wt%. The densities of gold [42], SWCNT [45], PH1000 [44], PVAc [12] used for calculating vol% respectively were 19.3, 1.3, 1.06, and 1.19 g/cm3. The density of MWCNT is 2 g/cm3 [46], which is different from that of SWCNT. Due to the difference, the contents of the samples containing MWCNT were not the same as those of SWCNT samples. For the sample 9, the weights of MWCNT, HAuCl4, PH1000, and PVAc were 0.0274 g, 0.1141 g, 0.2424 g, and 0.0049 g, respectively. The samples without gold were also prepared with SWCNT and MWCNT. In sample 8, 0.0641 g of CNT, 1.7420 g of PH1000, and 0.0177 g of PVAc were mixed, whereas 0.0733 g of MWCNT, 1.2951 g of PH1000, and 0.0132 g of PVAc were used in sample 7.

Electrical conductivity was obtained by a four-point probe method (current-voltage sweeping) and thermopower was acquired by measuring temperature differences and voltages across the samples at room temperature. Details can be found from our previous work [14]. The error bars were obtained from 2–4 measurements and uncertainties associated with dimensions (length, width, and thickness of the samples) and thermocouple reading. Errors were calculated with error propagation methods [1]. For electron microscopy analysis, the composites were cold-fractured by submerging the composites in liquid nitrogen for 5 min, and then the cross section of the composites was inspected.

AUTHOR CONTRIBUTIONS

Conceived and designed the experiments: CY KC. Performed the experiments: KC CY. Analyzed the data: KC CY. Contributed reagents/materials/analysis tools: KC CY. Wrote the paper: CY KC.

REFERENCES

1. Tans SJ, Verschueren ARM, Dekker C (1998) Room-temperature transistor based on a single carbon nanotube. Nature 393: 49–52.
2. Derycke V, Martel R, Appenzeller J, Avouris P (2001) Carbon nanotube inter- and intramolecular logic gates. Nano Lett 1: 453–456.
3. Liu XM, Romero HE, Gutierrez HR, Adu K, Eklund PC (2008) Transparent boron-doped carbon nanotube films. Nano Lett 8: 2613–2619. doi: 10.1021/

nl0729734

4. Hellstrom SL, Lee HW, Bao ZN (2009) Polymer-assisted direct deposition of uniform carbon nanotube bundle networks for high performance transparent electrodes. ACS Nano 3: 1423–1430. doi: 10.1021/nn9002456
5. Li J, Papadopoulos C, Xu JM, Moskovits M (1999) Highly-ordered carbon nanotube arrays for electronics applications. Appl Phys Lett 75: 367–369. doi: 10.1063/1.124377
6. Lee NS, Chung DS, Han IT, Kang JH, Choi YS, et al. (2001) Application of carbon nanotubes to field emission displays. Diam Relat Mat 10: 265–270. doi: 10.1063/1.124377
7. Kymakis E, Alexandrou I, Amaratunga GAJ (2003) High open-circuit voltage photovoltaic devices from carbon-nanotube-polymer composites. J Appl Phys 93: 1764–1768. doi: 10.1063/1.124377
8. Kongkanand A, Dominguez RM, Kamat PV (2007) Single wall carbon nanotube scaffolds for photoelectrochemical solar cells. Capture and transport of photogenerated electrons. Nano Lett 7: 676–680. doi: 10.1021/nl0627238
9. Ryu Y, Yu C (2011) The influence of incorporating organic molecules or inorganic nanoparticles on the optical and electrical properties of carbon nanotube films. Solid State Commun 151: 1932–1935. doi: 10.1016/j.ssc.2011.09.022
10. Meng CZ, Liu CH, Fan SS (2010) A promising approach to enhanced thermoelectric properties using carbon nanotube networks. Adv Mater 22: 535–539. doi: 10.1002/adma.200902221
11. Yao Q, Chen LD, Zhang WQ, Liufu SC, Chen XH (2010) Enhanced thermoelectric performance of single-walled carbon nanotubes/polyaniline hybrid nanocomposites. ACS Nano 4: 2445–2451. doi: 10.1021/nn1002562
12. Yu C, Kim YS, Kim D, Grunlan JC (2008) Thermoelectric behavior of segregated-network polymer nanocomposites. Nano Lett 8: 4428–4432. doi: 10.1021/nl802345s
13. Kim D, Kim Y, Choi K, Grunlan JC, Yu C (2010) Improved thermoelectric behavior of nanotube-filled polymer composites with poly(3,4-ethylenedioxythiophene) poly(styrenesulfonate). ACS Nano 4: 513–523. doi: 10.1021/nn9013577
14. Yu C, Choi K, Yin L, Grunlan JC (2011) Light-weight flexible carbon nanotube based organic composites with large thermoelectric power factors. ACS Nano 5: 7885–7892. doi: 10.1021/nn202868a
15. Yu C, Ryu Y, Yin L, Yang H (2011) Modulating electronic transport properties of carbon nanotubes yo improve the thermoelectric power factor via nanoparticle decoration. ACS Nano 5: 1297–1303. doi: 10.1021/nn102999h
16. Devadoss A, Dickinson C, Keyes TE, Forster RJ (2011) Electrochemiluminescent metallopolymer-nanoparticle composites: Nanoparticle size effects. Anal Chem 83: 2383–2387. doi: 10.1021/ac102697c
17. Podhajecka K, Dammer O, Pfleger J (2008) Electrical conductivity of poly(3-

octylthiophene)/Au nanocomposites. Macromol Symp 268: 72–76. doi: 10.1002/masy.200850815

18. Forster RJ, Keane L (2003) Nanoparticle-metallopolymer assemblies: charge percolation and redox properties. J Electroanal Chem 554: 345–354. doi: 10.1016/s0022-0728(03)00258-4
19. Ryu Y, Freeman D, Yu C (2011) High electrical conductivity and n-type thermopower from double-/single-wall carbon nanotubes by manipulating charge interactions between nanotubes and organic/inorganic nanomaterials. Carbon 49: 4745–4751. doi: 10.1016/s0022-0728(03)00258-4
20. http://en.wikipedia.org/wiki/Standard_electrode_potential_(data_page). Accessed 2012 Aug 20.
21. Choi HC, Shim M, Bangsaruntip S, Dai HJ (2002) Spontaneous reduction of metal ions on the sidewalls of carbon nanotubes. J Am Chem Soc 124: 9058–9059. doi: 10.1021/ja026824t
22. Kong BS, Geng JX, Jung HT (2009) Layer-by-layer assembly of graphene and gold nanoparticles by vacuum filtration and spontaneous reduction of gold ions. Chem Commun 2174–2176. doi: 10.1016/s0022-0728(03)00258-4
23. Kong BS, Jung DH, Oh SK, Han CS, Jung HT (2007) Single-walled carbon nanotube gold nanohybrids: Application in highly effective transparent and conductive films. J Phys Chem C 111: 8377–8382. doi: 10.1021/jp071297r
24. Yang SB, Kong BS, Kim DW, Baek YK, Jung HT (2010) Effect of Au doping and defects on the conductivity of single-walled carbon nanotube transparent conducting network films. J Phys Chem C 114: 9296–9300. doi: 10.1021/jp071297r
25. Shiraishi M, Ata M (2001) Work function of carbon nanotubes. Carbon 39: 1913–1917. doi: 10.1021/jp071297r
26. Sun JP, Zhang ZX, Hou SM, Zhang GM, Gu ZN, et al. (2002) Work function of single-walled carbon nanotubes determined by field emission microscopy. Appl Phys A-Mater Sci Process 75: 479–483. doi: 10.1021/jp071297r
27. Ago H, Kugler T, Cacialli F, Salaneck WR, Shaffer MSP, et al. (1999) Work functions and surface functional groups of multiwall carbon nanotubes. J Phys Chem B 103: 8116–8121. doi: 10.1021/jp071297r
28. Riviere JC (1966) The work function of gold. Appl Phys Lett 8: 172–173. doi: 10.1063/1.1754539
29. Sachtler WM, Dorgelo GJH, Holscher AA (1966) The work function of gold. Surf Sci 5: 221–229. doi: 10.1063/1.1754539
30. Ryu Y, Yin L, Yu C (2012) Dramatic electrical conductivity improvement of carbon nanotube networks by de-bundling and hole-doping with chlorosulfonic acid. J Mater Chem 22: 6959–6964. doi: 10.1063/1.1754539
31 Rowe DM (1995) CRC Handbook of Thermoelectrics. Boca Raton, Florida: CRC Press.
32. Kymakis E, Amaratunga GAJ (2006) Electrical properties of single-wall carbon nanotube-polymer composite films. J Appl Phys 99: 084302–084307. doi: 10.1063/1.1754539

33. Carroll DL, Czerw R, Webster S (2005) Polymer-nanotube composites for transparent, conducting thin films. Synth Met 155: 694–697. doi: 10.1016/j.synthmet.2005.08.031
34. Zhang B, Sun J, Katz HE, Fang F, Opila RL (2010) Promising thermoelectric properties of commercial PEDOT:PSS materials and their Bi(2)Te(3) powder composites. ACS Appl Mater Interfaces 2: 3170–3178. doi: 10.1021/am100654p
35. Reyes-Reyes M, Cruz-Cruz I, Lopez-Sandoval R (2010) Enhancement of the electrical conductivity in PEDOT:PSS films by the addition of dimethyl sulfate. J Phys Chem C 114: 20220–20224. doi: 10.1016/j.synthmet.2005.08.031
36. Kim JY, Jung JH, Lee DE, Joo J (2002) Enhancement of electrical conductivity of poly(3,4-ethylenedioxythiophene)/poly(4-styrenesulfonate)by a change of solvents. Synth Met 126: 311–316. doi: 10.1016/j.synthmet.2005.08.031
37. Hecht DS, Heintz AM, Lee R, Hu LB, Moore B, et al. (2011) High conductivity transparent carbon nanotube films deposited from superacid. Nanotechnology 22: 075201–075205. doi: 10.1088/0957-4484/22/7/075201
38. http://www.cheaptubesinc.com/MWNTs.htm. Accessed 2012 Aug 20.
39. http://www.cheaptubesinc.com/swnts.htm. Accessed 2012 Aug 20.
40. Geng HZ, Kim KK, Lee K, Kim GY, Choi HK, et al. (2007) Dependence of material quality on performance of flexible transparent conducting films with single-walled carbon nanotubes. Nano 2: 157–167. doi: 10.1142/s1793292007000532
41. Dai HJ, Wong EW, Lieber CM (1996) Probing electrical transport in nanomaterials: Conductivity of individual carbon nanotubes. Science 272: 523–526. doi: 10.1142/s1793292007000532
42. http://en.wikipedia.org/wiki/Gold. Accessed 2012 Aug 20.
43. http://www.unidym.com/files/Unidym_Product_Sheet_SWNT021810RevB.pdf. Accessed 2012 Aug 20.
44. Hu XJ, Jiang LN, Goodson KE (2004) Thermal conductance enhancement of particle-filled thermal interface materials using carbon nanotube inclusions. Itherm 1: 63–69.
45. Collins PG, Avouris P (2000) Nanotubes for electronics. Sci Am 283: 62–69. doi: 10.1038/scientificamerican1200-62
46. Park W, Choi K, Lafdi K, Yu C (2012) Influence of Nanomaterials in Polymer Composites on Thermal Conductivity. J Heat Trans-T ASME 134: 041302.

Chapter 3

BIOSAFETY OF NON-SURFACE MODIFIED CARBON NANOCAPSULES AS A POTENTIAL ALTERNATIVE TO CARBON NANOTUBES FOR DRUG DELIVERY PURPOSES

Alan C. L. Tang, Gan-Lin Hwang, Shih-Jung Tsai, Min-Yao Chang, Zack C. W. Tang, Meng-Da Tsai, Chwan-Yao Luo, Allan S. Hoffman,Patrick C. H. Hsieh Mail

Institute of Clinical Medicine, National Cheng Kung University & Hospital, Tainan, Taiwan

Nano-Powder and Thin Film Technology Center, Industrial Technology Research Institute, Tainan, Taiwan

Institute of Biomedical Engineering, National Cheng Kung University & Hospital, Tainan, Taiwan

Center for Micro/Nano Science and Technology, National Cheng Kung University & Hospital, Tainan, Taiwan

Department of Surgery, National Cheng Kung University & Hospital, Tainan, Taiwan

Department of Bioengineering, University of Washington, Seattle, Washington, United States of America

Institute of Biomedical Sciences, Academia Sinica, Taipei, Taiwan
Corresponding Author
Email: phsieh@mail.ncku.edu.tw

ABSTRACT

Carbon nanotubes (CNTs) have found wide success in circuitry, photovoltaics, and other applications. In contrast, several hurdles exist in using CNTs towards applications in drug delivery. Raw, non-modified CNTs are widely known for their toxicity. As such, many have attempted to reduce CNT toxicity for intravenous drug delivery purposes by post-process surface modification. Alternatively, a novel sphere-like carbon nanocapsule (CNC) developed by the arc-discharge method holds similar electric and thermal conductivities, as well as high strength. This study investigated the systemic toxicity and biocompatibility of different non-surface modified carbon nanomaterials in mice, including multi-walled carbon nanotubes (MWCNTs), single-walled carbon nanotubes (SWCNTs), carbon nanocapsules (CNCs), and C60 fullerene (C60). The retention of the nanomaterials and systemic effects after intravenous injections were studied.

Methodology and Principal Findings

MWCNTs, SWCNTs, CNCs, and C60 were injected intravenously into FVB mice and then sacrificed for tissue section examination. Inflammatory cytokine levels were evaluated with ELISA. Mice receiving injection of MWCNTs or SWCNTs at 50 μg/g b.w. died while C60 injected group survived at a 50% rate. Surprisingly, mortality rate of mice injected with CNCs was only at 10%. Tissue sections revealed that most carbon nanomaterials retained in the lung. Furthermore, serum and lung-tissue cytokine levels did not reveal any inflammatory response compared to those in mice receiving normal saline injection. Carbon nanocapsules are more biocompatible than other carbon nanomaterials and are more suitable for intravenous drug delivery. These results indicate potential biomedical use of non-surface modified carbon allotrope. Additionally, functionalization of the carbon nanocapsules could further enhance dispersion and biocompatibility for intravenous injection.

INTRODUCTION

The superior electrical and thermal conductivities, optical properties, and mechanical strength of carbon nanotubes (CNTs) and C60 fullerene (C60) make these nanomaterials ideal for use in structural supports, circuits, biosensors, batteries and solar cells [1], [2]. Different forms of fullerene have been envisioned as components of potential therapeutic devices in which they might act as tissue scaffolds 3, implants [4], biological microelectromechanical systems, biosensors, medical contrast agents, and drug delivery carriers [5]–[8]. Accordingly, the toxicology of CNTs has been widely investigated to understand the biological effects of these nanomaterials. Previous studies have demonstrated the in vivo toxicity and poor biocompatibility of multi-walled CNTs (MWCNTs), single-walled CNTs (SWCNTs) [9], [10] and C60 11 following inhalation 12, intratracheal instillation 13 or intraperitoneal injection [15]–[18].

Nanomaterials have been investigated as a technology to deliver therapeutic agents within the body with the ability to bypass tough biological barriers 19. Like most nanomaterials, the dimensions of CNTs are on the nanoscale, providing a high surface-area-to-volume ratio for efficient drug conjugation or encapsulation. Because great interest in using fullerenes for drug delivery has been generated, different forms of these carbon nanomaterials have been developed. To effectively use these nanomaterials for drug delivery, the biocompatibility and toxicity of these nanomaterials within biological systems must be fully characterized and understood [20]. Several reviews and studies have reported the toxicity of unmodified MWCNTs, SWCNTs, and C60 [9]–. The van der Waals forces on the surfaces of pristine CNTs cause hydrophobic interactions between CNTs, resulting in unwanted aggregation, agglomeration and wiring[9], [18], [20]. To avoid excessive surface interactions and to decrease toxicity, studies have opted to cut and extensively surface-modify CNTs for enhanced biocompatibility [4]–[10], [15]–[18], [20]. Despite so, overwhelming toxicological reports of CNTs have given rise to the consensus that these long and rigid CNTs are not suitable for in vivo applications [10]. Though surface modifications do in fact reduce toxicity to certain degree, the extensive act of functionalization and related modifications is simply masking the root cause of toxicity of CNTs, derived from the material's surface.

Recently, carbon nanocapsules (CNCs) have emerged as a novel carbon-based nanomaterial synthesized in a manner similar to that used for CNTs and C60 [21], [22], providing comparable chemical composition, and electrical, thermal, and mechanical characteristics. Following the footsteps of CNTs, CNCs have also found success in different applications, including transceiver modules and photovoltaics [23], [24], [25]. In contrast, biomedical applications using CNCs have not yet been attempted. A major difference, namely the aspect ratio, exists in the spherical geometry of CNCs, compared to long, tangling characteristics of CNTs. Intuitively, the low aspect ratio structure of CNCs are more dynamically suitable for in vivo delivery. Herein, we investigate the in vivo biocompatibility of non-modified CNCs, MWCNTs, SWCNTs, and C60 in mice, providing insight into advantages of using carbon nanocapsules for systemic drug delivery.

MATERIALS AND METHODS

Ethics Statement

All animal procedures were approved by the National Cheng Kung University Institutional Animal Care and Use Committee.

Carbon Nanomaterial Preparation

The CNCs were prepared as described previously [21], [22]. Briefly, an inert gas (helium, argon, or nitrogen), was introduced into an arc chamber containing a graphitic cathode and anode. A current was then introduced to the chamber that had sufficient voltage (10–30 V) for a carbon arc reaction to take place. The rate of the inert gas was controlled to approximately 60 to 90 cm3/min, and the chamber pressure was maintained between 1 and 2 atm. A pulse current was used (50–70 Hz; 50–500 A), and the deposit on the cathode was collected and passed through a 0.22 μm filter for purification. The deposits contained roughly 70% CNCs before purification and at least 95% after purification. MWCNTs were produced in a similar manner using the arc-discharge method under an argon atmosphere, as previously described [23]. A direct current electric field was applied, and deposits were collected from the cathode and purified. The deposits were roughly 50% pure and became more than 95% pure after purification. SWCNTs were purchased from SES research (Texas, USA) and C60 from Sigma Aldrich (Missouri, USA). All carbon nanomaterials were dispersed in 1 wt % polyvinyl alcohol (PVA) at 5–10 mg/ml. Immediately before injection, the nanomaterial dispersions were sonicated for 1 h (E60H, Elma Ultrasonics, Germany).

TEM Analysis

Carbon nanoparticles were dispersed in 1% PVA onto Formvar/carbon-coated 200 mesh copper grids (Ted Pella Inc, CA, USA) for TEM analysis using an H-7500 TEM (Hitachi, Japan). The samples were lyophilized for 24 h and imaged by an experienced technician.

Animal Protocols and Experiments

The investigation conformed to the Guide for the Care and Use of Laboratory Animals published by the US National Institutes of Health (NIH Publication No. 85–23, revised 1996). Adult FVB mice 8 to 12 weeks old and weighing 20 to 28 g were used in this study. The mice were anesthetized with sodium pentobarbital (50 mg/kg, i.p.). After the anesthetic had taken effect, the mice were injected with normal saline, 1% PVA, CNCs, C60, SWCNTs, or MWCNTs at a dose of 50, 25 or 12.5 µg/g bw via the tail vein, comparable to previous studies [26]–[28]. Normal saline and 1% PVA were injected at the same volume as the nanomaterials. Nanomaterials were injected at a diluted concentration so that each injection was approximately 200–250 µl (injection volume varied due to animal weight variations). A gauge was used to stop the bleeding of the tail, and the mice were allowed to recover from anesthesia in cages in a temperature- and climate-controlled environment with food and water. Mice were deeply anesthetized prior to sacrifice by cervical dislocation at different time points for organ harvesting and urine and blood collection. Blood collection was performed by cardiac puncture prior to cervical dislocation.

Mouse Survival Study

Mice were separated into three dosage groups with 11 or 12 mice in each group. The carbon nanomaterials were sonicated for 1 h prior to injection, and a total volume of 200–250 μl was injected into each mouse (injection volume varied due to animal weight variations). Nanomaterials were diluted with 1% PVA to maintain similar injection volumes for all three doses. After injection, mice were returned to their cages to recover from the anesthesia in a temperature- and climate-controlled environment. The mice were monitored closely for the first 6 hours, at 12 hours, and then every day following the first day. The time of death of each mouse was recorded, and at the end of 7 d, the mice were sacrificed, and their organs were harvested for tissue sectioning.

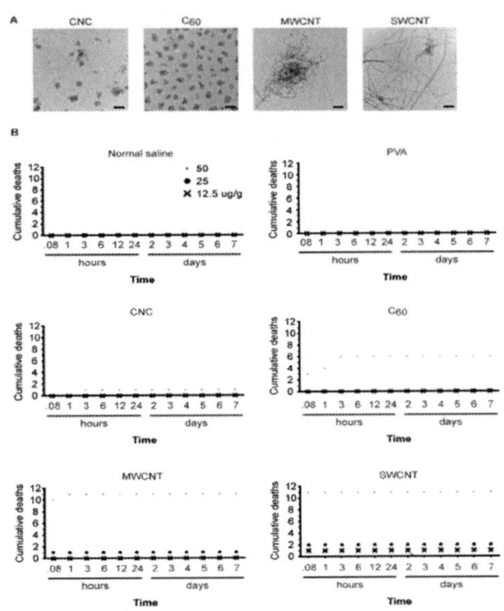

Figure 1. Mouse survival curves after carbon nanomaterial injection.

(A) TEM analysis of carbon nanocapsules (CNCs), C60 fullerene (C60), multi-walled carbon nanotubes (MWCNTs), and single-walled carbon nanotubes (SWCNTs) dispersed in 1% polyvinyl alcohol (PVA). SWCNTs formed large networks, and MWCNTs aggregated compactly. CNCs were well dispersed in PVA, while C60 aggregated to size as large as CNCs. The scale bar is 100 nm. (B) Cumulative deaths of mice intravenously injected with different doses of carbon nanomaterials. SWCNTs and MWCNTs had the highest toxicity, which was dose dependent, decreasing as the dose of the carbon nanomaterials decreased. No mortality was observed among the CNC-treated mice at 25 μg/g b.w. n = 12 for CNC, and C60 injected mice. n = 11 for NS, PVA, MWCNT, and SWCNT injected mice. Red square, 50 μg/g; black dot, 25 μg/g; black cross, 12.5 μg/g.

Tissue Sections and Nanomaterial Retention Quantification

Organs were harvested 6 h or 7 d after injection with the nanomaterials, washed in PBS and fixed in 4% paraformaldehyde at 4°C overnight. The organs were then stored in 70% ethanol prior to paraffin embedding. Sections were stained with hematoxylin and imaged using an Axio Scope A1 imaging system (Carl Zeiss). For the nanomaterial retention study, at least 2 tissue sections from each lung were used and were left unstained to reduce background and false positive signals. Whole tissue sections were imaged by HistoFAXS (TissueGnostics, Austria) at a 200x final magnification and were analyzed with HistoQuest (TissueGnostics, Austria) for automated structure detection, automatic color separation, and quantification.

Inflammatory Cytokine Study

Blood was collected from mice 6 hours post-injection and was allowed to sit at room temperature for at least 1 h. Lipopolysaccharide (LPS) injected intravenously served as positive control (5 mg/kg, sigma). Samples were centrifuged at 1500 g for 10 minutes to obtain serum. Serum samples were then analyzed using ELISA kits for mouse IL-1β and mouse IL-6 for the detection of cytokines. These assays were performed according to the manufacturer's instructions (AssayPro, USA). Normal saline was used as a negative control, and lipopolysaccharide was used as a positive control. Lung tissues were also collected 6 hours post-injection and homogenized in 500 μl of lysis buffer containing protease inhibitors. Homogenized samples were centrifuged at 14,000 RPM for 20 minutes to remove debris. The supernatants were analyzed using ELISA kits for mouse IL-1β and mouse IL-6 for the detection of cytokines. Cytokine levels were normalized to the total protein level determined by the BCA assay

(Pierce, USA).

Statistical Analysis

All data are presented as the mean±sem. Data were analyzed by one-way ANOVA followed by Tukey's post hoc test using Prism 5 (GraphPad, USA). A value of $P<0.05$ was considered statistically significant.

RESULTS

Physical characteristics of carbon nanomaterials

Non-modified CNTs, whether single-walled (SWCNT) or multi-walled (MWCNT), form networks and aggregates even when dispersed in a surfactant such as PVA (Fig. 1a). The tube diameter of MWCNT were approximately 25 nm measured from TEM images. SWCNT diameters ranged from 2 nm to 25 nm (Table S1). C60, though much more uniform and dispersed than CNTs, still aggregated, forming 100 nm in diameter clusters. C60 molecules have a very low solubility and an extremely high density. Therefore, these nanoparticles settle within minutes even when dispersed in PVA after sonication and mixing (Figure S1). CNCs were much more uniformly dispersed and each particle was approximately 50 nm in diameter (Fig. 1a).

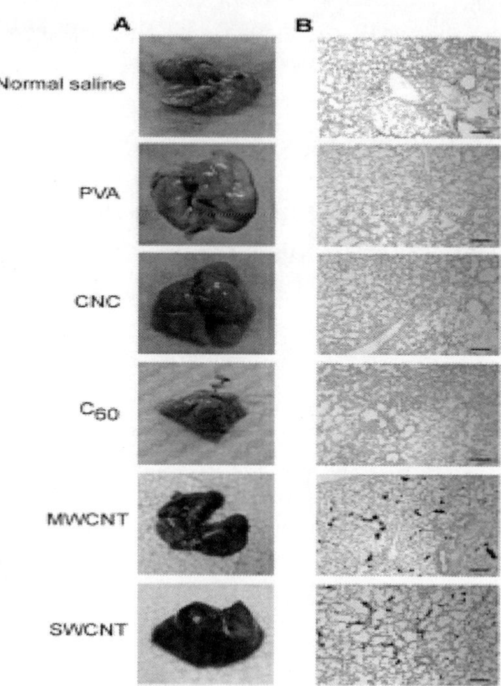

Figure 2. Lung tissues and lung tissue sections after carbon nanomaterial injection.

(A) Excised lungs 10 min after mice were injected with 50 μg/g b.w. of different carbon nanomaterials. (B) Lung tissue sections 10 min after mice were injected with 50 μg/g b.w. of different carbon nanomaterials. Mice receiving C60 fullerene (C60), multi-walled carbon nanotubes (MWCNTs), and single-walled carbon nanotubes (SWCNTs) died within 10 minutes, and only mice in the carbon nanocapsule (CNC), normal saline, and polyvinyl alcohol (PVA) groups had to be sacrificed. Tissue sections were stained with hematoxylin. The scale bar is 400 μm. doi:10.1371/journal.pone.0032893.g002,

In vivo toxicity of carbon nanomaterials and cause of death

To study the in vivo toxicity of carbon nanomaterials, different carbon nanomaterials were intravenously injected into mice at three different doses. Strikingly, none of the mice receiving 50 μg/g b.w. of either MWCNTs or SWCNTs survived (n = 11 for each). By contrast, mice injected with CNCs had a 91.7% survival rate (n = 12), while half of the mice injected with C60 died (6 out of n = 12; Fig. 1b). The toxicities of the nanomaterials were dose dependent, as shown by the survival curves for the three different doses (Fig. 1b). Interestingly, all of the mice injected with 25 and 12.5 μg/g b.w. of CNCs survived, while some of the mice injected with these doses of MWCNTs or SWCNTs died (Fig. 1b).

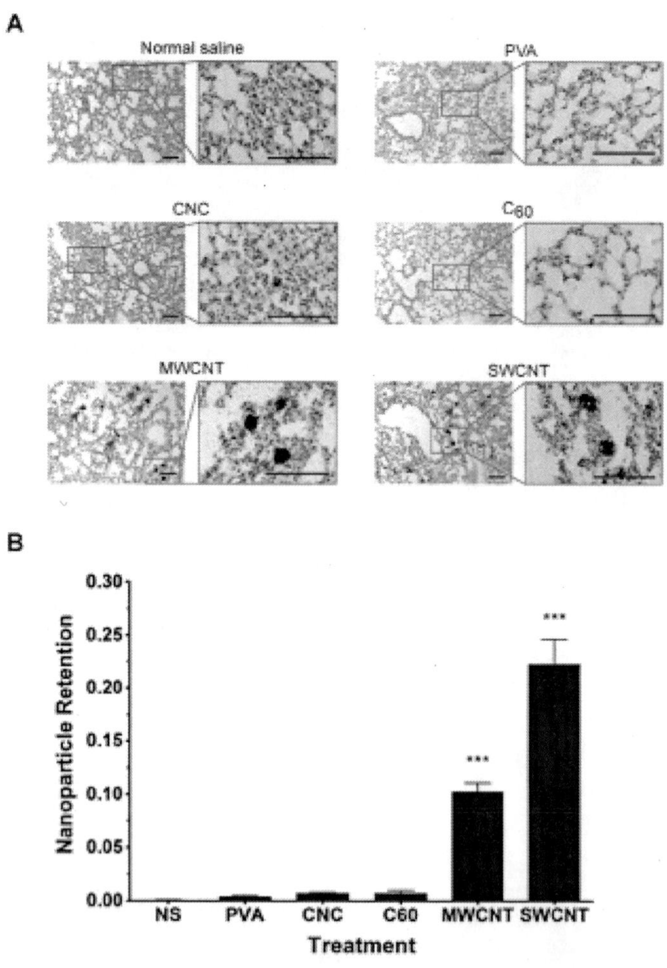

Figure 3. Carbon nanomaterial retention in the lungs.

(A) Lung tissue sections of mice 7 days after intravenous injection with carbon nanomaterials at 25 μg/g b.w. (B) Automatic carbon nanomaterial retention quantification in the lungs. High-magnification images (red-bordered images) show large carbon nanotube aggregates blocking the blood vessels of the lungs (arrows). SWCNTs and MWCNTs were retained in the lungs at much higher rates compared to CNCs or C60. Tissue sections were stained with hematoxylin. Scale bar = 100 μm. ***P<0.0001 compared to CNCs and C60, n = 4 in all groups. NS, normal saline; PVA, polyvinyl alcohol; CNCs, carbon nanocapsules; C60, C60 fullerene; MWCNTs, multi-walled carbon nanotubes; SWCNTs, single-walled carbon nanotubes.

Postmortem inspections of the mice receiving 50 μg/g doses revealed that CNTs were clearly visible in the lungs (Fig. 2a). As expected, the MWCNT- and SWCNT-injected groups had the darkest lungs, which were fully covered with black spots. Lungs from CNC- and C60-injected mice generally exhibited a pink hue similar to that of the normal saline and 1% PVA in deionized H2O (PVA) groups, both of which served as controls (Fig. 2a). Tissue sections of these lungs further revealed that a large surface area of MWCNT and SWCNT lungs was occupied by CNTs (Fig. 2b).

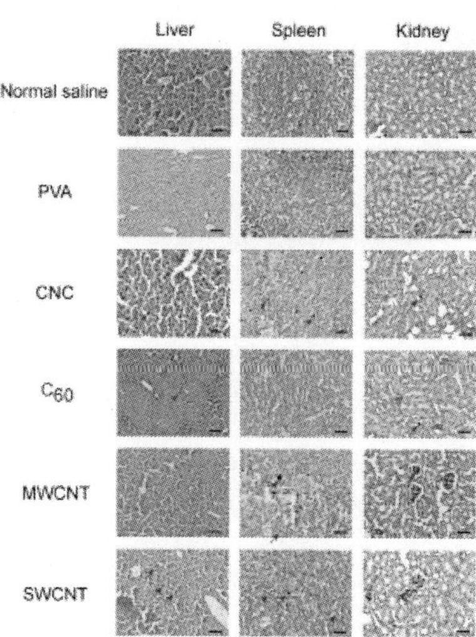

Figure 4. Carbon nanomaterial retention in vital organs.

Liver, spleen, and kidney tissue sections of mice 7 days after intravenous injection with carbon nanomaterials at 25 μg/g b.w. Carbon nanomaterials are indicated by arrows. Tissue sections were stained with hematoxylin. The scale bar is 50 μm. PVA, polyvinyl alcohol; CNCs, carbon nanocapsules; C60, C60 fullerene; MWCNTs, multi-walled carbon nanotubes; SWCNTs, single-walled carbon nanotubes.

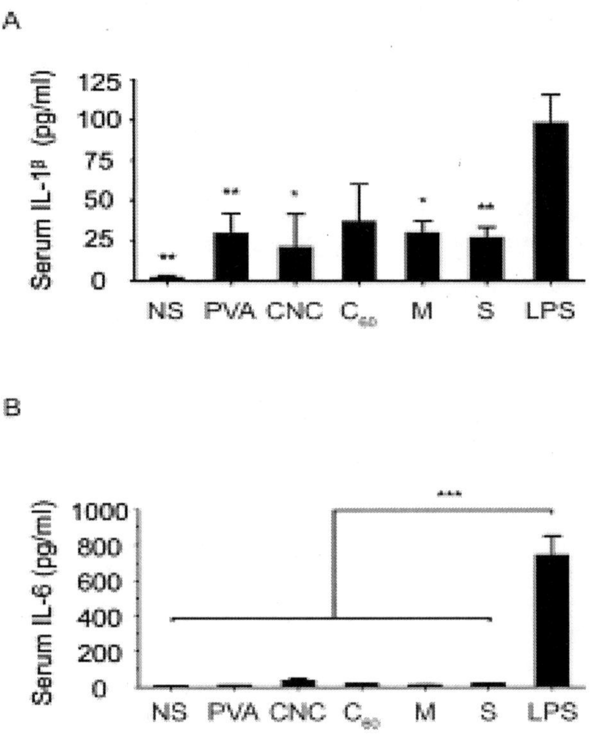

Figure 5. Systemic inflammatory cytokine level in mice.

Serum and lung tissue IL-1β and IL-6 levels 6 hours post-injection with carbon nanomaterials. There is no significant difference between all groups (except LPS). n = 5, P = 0.0029 (serum, IL-1β); n = 5, P = 0.0001 (serum, IL-6); n = 4. $^*P<0.05$, $^{**}P<0.01$, $^{***}P<0.001$ significantly different compared with the LPS group. NS, normal saline; PVA, polyvinyl alcohol; CNCs, carbon nanocapsules; C60, C60 fullerene; M, multi-walled carbon nanotubes; S, single-walled carbon nanotubes.

Biodistribution and Retention of Carbon Nanomaterials

To quantify and compare the retention of the carbon nanomaterials in the lungs, surviving mice from the 25 μg/g dose injections were sacrificed on day 7, and the lungs were collected for tissue section analysis. Similar to injections of 50 μg/g, lung tissue sections from mice injected with 25 μg/g showed a similar trend in the lung retention of the nanomaterials. MWCNTs and SWCNTs were widely distributed and accumulated in the lungs, while CNCs and C60 were scarce (Fig. 3a). Automated microscopic whole tissue section analysis revealed that MWCNTs retained in the lungs by more than a factor of 14 compared to CNCs or C60, while SWCNTs retained by a factor of more than 30 (Fig. 3b). Because CNTs formed larger aggregates in the range of 200–1000 nm (Fig. 1), more blood vessels were observed to be have been clogged by MWCNTs and SWCNTs (Fig. 3a). By contrast, CNCs and C60 did not form aggregates larger than 200 nm; therefore, they passed through to other organs including the liver, spleen, and kidney, or cleared through the renal system (Fig. 4, Figure S2). SWCNTs were lethal even at the dose of 12.5 μg/g following systemic injection, and retention was found to be twice that of MWCNTs in the lungs.

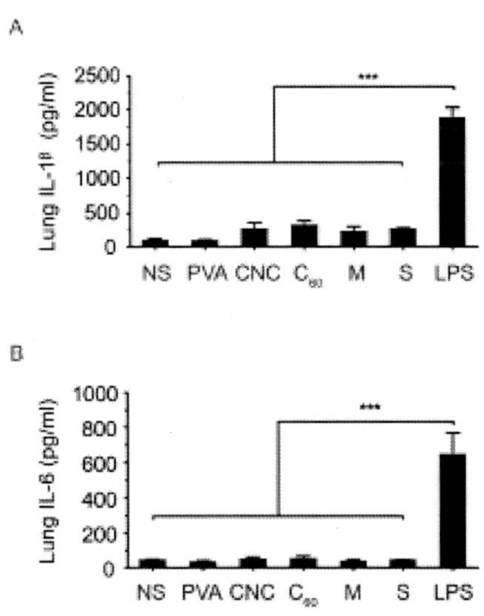

Figure 6. Lung tissue inflammatory cytokine level in mice.

Lung tissue IL-1β and IL-6 levels 6 hours post-injection with carbon nanomaterials. There is no significant difference between all groups (except LPS). Lung-tissue cytokine levels were normalized to the total protein level determined using a BCA kit (Pierce, USA). n = 5, ***P = 0.0001 (lung tissue, IL-1β); n = 4, P = 0.0001 (lung tissue, IL-6), compared to the LPS group. NS, normal saline; PVA, polyvinyl alcohol; CNCs, carbon nanocapsules; C60, C60 fullerene; M, multi-walled carbon nanotubes; S, single-walled carbon nanotubes.

Systemic inflammatory response after IV injection of carbon nanomaterials

Although CNTs and nanoparticles have the potential to be used for drug delivery applications, the foreign body reaction of nanomaterials is a concern. Nanotoxicity arises from the inflammatory responses to foreign bodies, cellular uptake and inflammatory cytokine production, among many other acute responses [29], [30]. Interleukin–1 beta (IL-1β) and interleukin–6 (IL-6) are two important inflammatory cytokines induced during inflammation that mediate the inflammatory response. To examine the acute systemic response after injecting the nanomaterials, serum was collected from the surviving mice for an ELISA analysis. Lipopolysaccharide (LPS) served as a positive control for inducing systemic inflammatory cytokines. Serum from all groups of injections showed no significant difference with each other in IL-1β and IL-6 expression levels (Fig. 5). Furthermore, the cytokine levels in the lung tissue samples also showed the same result (Fig. 6). LPS injected samples, both serum and lung tissue, were significantly were significantly higher in both IL-1β and IL-6. Pellets from lung-tissue preparation for cytokine detection revealed consistent results from tissue sections (Figure S3). MWCNT and SWCNT lung homogenates were extremely dark while CNC and C60 groups were much lighter in color.

DISCUSSION

Although nanoscopic in feature size, pristine CNT surfaces strongly attract each other through van der Waals forces, causing aggregation and network formation [5], [9], [18], [20]. This non-dispersing interaction prevents CNT from being an ideal tool for drug delivery through intravenous injection [18], [20]. Consistent with previous studies, we found that CNTs were prone to aggregate and easily bundled on itself [10]. To overcome the challenging aggregate-forming surface properties of CNTs, previous studies have put effort in cutting, functionalizing, and surface modifying CNTs [4]–[10], [15]–[18], [20]. However, surface modification is rather masking an existing flaw of CNT toxicity instead of actually eliminating this shortcoming. Most studies using CNTs for drug delivery purposes require cutting to reduce the overall length. However, in this form, CNTs are still too long and rigid drug delivery purposes [10]. Furthermore, different fabrication protocols introduce defects as part of the process, whether intentionally or as a side effect [31], [32]. These structural defects have been linked to causing acute lung toxicity, genotoxicity, and inflammatory responses [33]–[36]. Recently, CNCs have been produced using a method similar to that used for preparing CNTs using a pulsed plasma arc discharge method [21], [22]. Much like CNTs, CNCs have high electrical conductivity, thermal conductivity, strength, and surface-area-to-volume ratio [37]. However, CNCs differ from CNTs in that CNCs inherently have a much lower aspect ratio at around 1.5. Furthermore, CNCs are uniformly synthesized nanoparticles ranging from 40–60 nm while highly dense C60 form clusters up to 100 nm. Due to the previously mentioned properties, CNCs, unlike CNTs and C60, lack aggregating properties, which are much more favorably biocompatible for drug delivery purposes.

In our study, CNTs in the lungs formed aggregates of approximately the same diameter as the blood vessels and were trapped in a manner similar to a pulmonary embolism, clogging the blood vessels. Thus, the main cause of death of the mice injected with high dose CNTs is attributed to the mechanical obstruction of the blood vessels in the lungs, possibly leading to acute heart failure. By contrast, very few carbon nanomaterials were found in the CNC or C60 lung tissue sections. The CNC and C60 nanoparticles that remained in the lungs were extremely small, and most of these small nanoparticles may have escaped the highly vascularized lungs and traveled to other organs, while the aggregated MWCNTs and SWCNTs were easily trapped in the lungs. Consistent with previous reports, large aggregates of CNTs are the main concern for the biosafety of this material [20]. What remains unclear, however, is why there was high toxicity after the C60 injections despite the high clearance rate and the lack of an immune response. Lung and other vital organ tissue sections of the C60 group did not show any retention at any dose. Although C60 was absent from our tissue sections, we cannot rule out C60 retention due to the extreme small dimensions of this nanoparticle. Indeed, a recent study quantified C60 retention following intravenous injection in rats using liquid chromatography [38]. Their results show high acute retention of the particles in the vital organs, particularly the lung one day following injection, which decreased over the course of 4 weeks. Together with other studies that have reported C60 toxicity, it is reasonable that C60 fuller is lethal following intravenous injection.

The inflammatory response was consistent with other studies that CNTs or other forms of fullerene did not elicit an inflammatory response in any tissue, although the nanomaterials were distributed throughout the animal [20]. Both systemic and local tissue studies showed that cytokine levels were at the same level as normal saline treated groups at the acute phase. LPS groups were significantly higher in both IL-1β and IL-6 levels. Long-term studies are required to further understand whether these nanomaterials trigger chronic inflammation. However, similar to our findings, a recent study by Burke et al provided similar insight into the toxicity profile of CNTs [39]. MWCNTs injected at similar dosage were lethal in the acute stage, mainly by obstructing blood vessels in the lung. Furthermore, non-functionalized MWCNTs increased vWF and D-dimer levels following systemic injection, and reduced platelet count. Additionally, functionalization is effective in attenuating coagulation effects of MWCNTs.

To our knowledge, there has not been a single study that had comprehensively investigated the in vivo effect of different carbon nanomaterials injected intravenously. Most past studies aimed to characterize production plant safety by studying the pulmonary toxicity of carbon nanomaterials encountered through intratracheal instillation [9]. In addition, we included a novel carbon allotrope, CNC in this study. Here, we investigated the in vivo toxicity of raw, non-modified carbon nanomaterials delivered intravenously, hoping to gain an understanding of the dynamics of these nanomaterials in vivo. Nanotechnology has offered a wealth of possibilities for enhanced drug therapy to deliver therapeutic treatment. However, nanotoxicity must be well characterized before this technology can be used safely and effectively. Our results show that although certain properties of CNTs have allowed propelled this material to succeed in applications in sensors, circuitry, and structural components, the aggregating property of CNTs inhibit safe usage in drug delivery.

In this preliminary study, CNCs have been shown to be more biocompatible following intravenous injection and may emerge in the future for drug delivery purposes. These CNCs have already been functionalized to further enhance dispersion rates. In the future, these functionalized CNCs represent a novel carbon allotrope as a solution to the aggregating issue of CNTs, providing an alternate research opportunity towards drug delivery. Using current established methods, crosslinkers can be employed to conjugate antibodies, proteins, peptides, or small molecules onto functionalized CNCs for drug delivery purposes. We envision CNCs as a potential alternative to CNTs in the application of intravenous drug delivery.

ACKNOWLEDGMENTS

We thank the Department of Chemistry of NCKU for TEM imaging assistance and Professor Mathew O'Donnell of the University of Washington, Seattle, for his comments and discussion on the subject. We also thank Pei-Yu Lee of NCKU for TissueGnostics tissue section image acquisitions.

AUTHOR CONTRIBUTIONS

Conceived and designed the experiments: ACLT ZCWT PCHH. Analyzed the data: ACLT GLH ZCWT MDT CYL ASH PCHH. Wrote the paper: ACLT. Performed the animal experiments: ACLT MYC ZCWT. Performed the TEM analysis: MYC ZCWT. Performed all ELISA experiments and tissue staining: ACLT. Prepared the MWCNTs and CNCs: GLH SJT. Read and revised the manuscript: SJT CYL ASH. Discussed the results and approved the final version of the manuscript: ACLT GLH SJT MYC ZCWT MDT CYL ASH PCHH.

REFERENCES

1. Baughman RH, Zakhidov AA, de Heer WA (2002) Carbon nanotubes - the route toward applications. Science 297: 787–792.
2. Li C, Chen Y, Wang Y, Iqbal Z, Chhowalla M, et al. (2007) A fullerene-single wall carbon nanotube complex for polymer bulk heterojunction photovoltaic cells. J Mater Chem 17: 2406–2411.
3. Harrison BS, Atala A (2007) Carbon nanotube applications for tissue engineering. Biomaterials 28: 344–353.
4. Saito N, Usui Y, Aoki K, Narita N, Shimiza M, et al. (2009) Carbon nanotubes: biomaterial applications. Chem Soc Rev 38: 1897–1903.
5. Zhang Y, Bai Y, Yan B (2010) Functionalized carbon nanotubes for potential medicinal applications. Drug Discov. Today 15: 428–435.
6. Cheng W, Pontoriero OT, Chen AM, He H (2010) DNA and carbon nanotubes as medicine. Adv Drug Deliver Rev 62: 633–649.
7. Podesta J, Al-Jamal KT, Herrero MA, Tian B, Ali-Boucetta H, et al. (2009) Antitumor Activity and Prolonged Survival by Carbon-Nanotube-Mediated Therapeutic siRNA in a human lung xenograft Model Small 5: 1176–1185.
8. Bhirde AA, Patel V, Gavard J, Zhang G, Sousa AA, et al. (2009) Targeted killing of cancer cells in vivo and in vitro with EGF-directed carbon nanotube-based drug delivery. ACS nano 3: 307–316.
9. Lacerda L, Bianco A, Prato M, Kostarelos K (2006) Carbon nanotubes as nanomedicines: from toxicology to pharmacology. Adv Drug Deliv Rev 58: 1460–1470.
10. Kostarelos K (2008) The long and short of carbon nanotube toxicity. Nature Biotechnol 26: 774–776.
11. Baker GL, Gupta A, Clark ML, Valenzuela BR, Staska LM, et al. (2008) Inhalation toxicity and lung toxicokinetics of C60 fullerene nanoparticles and microparticles. Toxicol Sci 101: 122–131.
12. Ryman-Rasmussen JP, Cesta MF, Brody AR, Shipley-Phillips JK, Everitt JI, et al. (2009) Inhaled carbon nanotubes reach the subpleural tissue in mice. Nature Nanotech 4: 747–751.
13. Muller J, Huaux F, Fonseca A, Nagy JB, Moreau N, et al. (2008) Structural defects play a major role in the acute lung toxicity of multiwall carbon nanotubes: toxicological aspects. Chem Res Toxicol 21: 1698–1705.
14. Xu JY, Li QN, Li JG, Ran TC, Wu SW, et al. (2007) Biodistribution of 99mTc-C60 (OH)xin Sprague-Dawley rats after intratracheal instillation. Carbon 45: 1865–1870.
15. Deng X, Jia G, Wang H, Sun H, Wang X, et al. (2007) Translocation and fate of multi-walled carbon nanotubes in vivo. Carbon 45: 1419–1424.
16. Poland CA, Duffin R, Kinloch I, Maynard A, Wallace WAH, et al. (2008) Carbon nanotubes introduced into the abdominal cavity show asbestos-like pathogenicity in a pilot study. Nature Nanotech 3: 423–428.

17. Yang W, Thordarson P, Gooding JJ, Ringer SP, Braet F (2007) Carbon nanotubes for biological and biomedical applications. Nanotechnology 18: 1–12.
18. Kostarelos K, Bianco A, Prato M (2009) Promises, facts, and challenges for carbon nanotubes in imaging and therapeutics. Nature Nanotech 4: 627–633.
19. LaVan DA, McGuire T, Langer R (2003) Small-scale systems for in vivo drug delivery. Nature Biotechnol 21, 1184–1191:
20. Aillon KL, Xie Y, El-Gendy N, Berkland CJ, Forrest ML (2009) Effects of nanomaterial physicochemical properties on in vivo toxicity. Adv Drug Deliv Rev 61, 45–466:
21. Hwang GL (2002) US patent 7156958:
22. Hwang GL, Hwang KC, Shieh YT, Lin SJ (2003) Preparation of carbon nanotube encapsulated copper nanowires and their use as a reinforcement for Y-Ba-Cu-O superconductors. Chem Mater 15: 1353–1357.
23. Cheng WH, Hung WC, Lee CH, Hwang GL, Jou WS, et al. (2004) Low-cost and low-electromagnetic-interference packaging of optical transceiver modules. J Lightw Technol 22: 2177–2183.
24. Hung HC, Hwang GL, Chen HL, Lee YD (2006) Immobilization of TiO2 nanoparticles on carbon nanocapsules for photovoltaic applications. Thin Solid Films 511–512: 203–207.
25. Hung HC, Hwang GL, Chen HL, Lee YD (2006) Immobilization of TiO2 nanoparticles on Fe filled carbon nanocapsules for photocatalytic applications. Thin Solid Films 515: 1033–1037.
26. Singh R, Pantarotto D, Lacerda L, Pastorin G, Klumpp C, et al. (2006) Tissue biodistribution and blood clearance rates of intravenously administered carbon nanotube radiotracers. Proc Natl Acad Sci U S A. 103: 3357–3362.
27. Zhang D, Deng X, Ji Z, Shen X, Dong L, et al. (2010) Long-term hepatotoxicity of polyethylene-glycol functionalized multi-walled carbon nanotubes in mice. Nanotechnology 21: 1–10.
28. Yang K, Zhang S, Zhang G, Sun X, Lee ST, et al. (2010) Graphene in mice: ultrahigh in vivo tumor uptake and efficient photothermal therapy. Nano Lett 10: 3318–3323.
29. Dobrovolskaia MA, Germolec DR, Weaver JL (2009) Evaluation of nanoparticle immunotoxicity. Nature Nanotech 4: 411–414.
30. Dobrovolskaia MA, McNeil SE (2007) Immunological properties of engineered nanomaterials. (2007).Nature Nanotech 2: 469–478.
31. Tong DG, Luo YY, Chu W, Guo YC, Tian W (2010) Cutting of carbon nanotubes via solution plasma processing. Plasma Chem Plasma Process 30: 897–905.
32. Jeong SH, Lee OJ, Lee KH (2002) Preparation of aligned carbon nanotubes with prescribed dimensions: Template synthesis and sonication cutting approach. Chem Mater 14: 1859–1862.
33. Muller J, Huaux F, Fonseca A, Nagy JB, Moreau N, et al. (2008) Structural defects play a major role in the acute lung toxicity of multiwall carbon

nanotubes: toxicological aspects. Chem Res Toxicol 21: 1698–705.

34. Fenoglio I, Greco G, Tomatis M, Muller J, Raymundo-Piñero E, et al. (2008) Structural defects play a major role in the acute lung toxicity of multiwall carbon nanotubes: physicochemical aspects. Chem Res Toxicol 21: 1690–7.

35. Sato Y, Ootsubo M, Yamamoto G, Van Lier G, Terrones M, et al. (2008) Super-robust, lightweight, conducting carbon nanotube blocks cross-linked by de-fluorination. ACS Nano 2008 2: 348–56.

36. Shvedova AA, Kisin ER, Porter D, Schulte P, Kagan VE, et al. (2009) Mechanisms of pulmonary toxicity and medical applications of carbon nanotubes: Two faces of Janus? Pharmacol Ther 121: 192–204.

37. Su TT (2006) Commercialization of Nanotechnology - Taiwan Experiences. Presented at: IEEE Conference: Emerging technologies - Nanoelectronics. Meritus Mandarin Hotel, Singapore, 10 January–13 January.

38. Kubota R, Tahara M, Shimizu K, Sugimoto N, Hirose A, et al. (2011) Time-dependent variation in the biodistribution of C60 in rats determined by liquid chromatography-tandem mass spectrometry. Toxicol Lett 206: 172–177.

39. Burke AR, Singh RN, Carroll DL, Owen JD, Kock ND, et al. (2011) Determinants of the thrombogenic potential of multiwalled carbon nanotubes. Biomaterials 32: 5970–5978.

Chapter 4

PLGA-CARBON NANOTUBE CONJUGATES FOR INTERCELLULAR DELIVERY OF CASPASE-3 INTO OSTEOSARCOMA CELLS

Qingsu Cheng, Marc-Olivier Blais, Greg Harris, and Ehsan Jabbarzadeh

Department of Biomedical Engineering, University of South Carolina, Columbia, South Carolina, United States of America

Department of Chemical Engineering, University of South Carolina, Columbia, South Carolina, United States of America

Department of Orthopaedic Surgery, University of South Carolina, Columbia, South Carolina, United States of America

ABSTRACT

Cancer has arisen to be of the most prominent health care issues across the world in recent years. Doctors have used physiological intervention as well as chemical and radioactive therapeutics to treat cancer thus far. As an alternative to current methods, gene delivery systems with high efficiency, specificity, and safety that can reduce side effects such as necrosis of tissue are under development. Although viral vectors are highly efficient, concerns have arisen from the fact that viral vectors are sourced from lethal diseases. With this in mind, rod shaped nano-materials such as carbon nanotubes (CNTs) have become an attractive option for drug delivery due to the enhanced permeability and retention effect in tumors as well as the ability to penetrate the cell membrane.

Here, we successfully engineered poly (lactic-co-glycolic) (PLGA) functionalized CNTs to reduce toxicity concerns, provide attachment sites for pro-apoptotic protein caspase-3 (CP3), and tune the temporal release profile of CP3 within bone cancer cells. Our results showed that CP3 was able to attach to functionalized CNTs, forming CNT-PLGA-CP3 conjugates. We show this conjugate can efficiently transduce cells at dosages as low as 0.05 µg/ml and suppress cell proliferation up to a week with no further treatments. These results are essential to showing the capabilities of PLGA functionalized CNTs as a non-viral vector gene delivery technique to tune cell fate.

INTRODUCTION

Cancer is a principal concern in today's health care treatment with more than 10 million cases each year [1]. Oncologists generally utilize physiological intervention as well as chemical and radioactive therapeutics to treat different forms of cancer. The drawbacks to the current therapeutics are the removal or necrosis of healthy tissue in addition to the tumor tissue. These side effects are one of the major concerns for traditional cancer treatments. To this end, targeted drug delivery has paved another avenue in potential cancer therapy in recent years.

The Food and Drug Administration (FDA) approved clinical gene therapy in the 1990s [2] and in experimental and clinical stages has since demonstrated the promise of drug delivery in curing human diseases. Retrovirus [3], adenovirus [4] and adeno-associate [5] viral vectors are the major types of gene vectors used in research today. Although viral vectors are efficient in both delivery and transduction of cells, there are concerns that these viruses are sourced from lethal diseases such as human immunodeficiency virus (HIV) and human T-cell lymphotropic virus (HTLV). A safe, inexpensive, and effective vector has yet to be fabricated for use in gene therapy. Therefore, developing these safe and efficient delivery systems in a controlled manner is a major focus and challenge in research [6].

The enhanced permeability and retention (EPR) effect, which is inherent to tumor biology, can allow nano-sized drug carriers to accumulate, retain, and release drugs due to the fact that tumors have leaky blood vessels and poor lymphatic drainage [7]. In further researching nano-sized drug carriers, scientists have begun widening the window into the dark room of cancer therapy. The use of nano-sized drug carriers has a few key advantages including the protection of delicate drugs, enhanced absorption in selective tissues, controlled drug distribution profile, and enhanced intracellular penetration [8]. Thus, researchers have taken initiative and developed several methods to fabricate nano-sized drug delivery carriers.

Polymers and lipids have garnered the most attention thus far as materials for drug delivery. Synthetic polymers such as poly (ethylene glycol) (PEG)[9], poly (lactic acid) (PLA) [10] and PLGA [11] as well as natural polymers such as chitosan[12], collagen [13], gelatin [14] or lipids [15] can be fabricated as nanoparticles, liposomes, and micelles to deliver molecules by either chemical modification or physical absorption. Drugs are able to be released in a controlled manner due to either surface and bulk degradation or phase transition principles. However, there are concerns due to immune responses dealing with the heterogeneity of the materials used as well as the low transfection efficiency and specificity. In noting this, we sought to develop a highly efficient non-viral vector drug delivery system utilizing CNTs.

CNTs [16], silicon nanowires [17], gallium nanotubes [18], boron nitride nanotubes [19], titanium oxide nanotubes [20] and zinc oxide nano-rods [21] have received an overwhelming amount of support and enthusiasm in biomedical research. CNTs are made of a layer of grapheme [22] and have been used to shuttle several different biological molecules, ranging from small drug molecules [23] to biomacromolecules such as proteins [24], DNA [25,26] and RNA [27] into different types of cells via endocytosis efficiently [28]. Ricin A is an example that has been delivered into numerous cell lines through conjugation of CNTs intended to induce cell death [29].

In addition to cell proliferation inhibition and induction of apoptosis, DNA plasmids were also delivered into cells and enhanced desired gene expression. CNTs are able to achieve this penetration of the cells and present the opportunity to transport biological cargo across the cell membrane due to an extremely high aspect ratio. However, health concerns have hampered the practical application of using inorganic CNTs in biological applications thus far. Toxicity issues still raise doubts about the practical applications of CNTs due to the intrinsic toxicity caused by their high surface area and hydrophobicity [30]. However, toxicity concerns are lessened in cancer therapy due to the EPR effect which accumulates nanoparticles in only tumor tissue limiting the potential toxicity of CNTs to strictly the cancer cells and the fact that CNTs at small amounts or functionalized with other molecules have shown to evoke minimum toxicity to normal cell lines [31,32]. Therefore, functionalization of CNTs can prove to be a successful path to protein and gene therapy. For instance, Liu et al. have been able to successfully deliver siRNA with phospholipid PEG (PL-PEG) functionalized CNTs into tumor cells and tissue to inhibit tumor cell proliferation, tissue ingrowth, and CXCR4 expressions both*ex vivo* and *in vivo* [33–35]. siRNA was released by breaking the S-S bond, which attached the siRNA to PL-PEG. Other pristine CNT based delivery vectors are able to rely on diffusion to unload biological cargoes[24–27]. Although the cellular uptake mechanism may differ depending on the functionalization and size of CNTs[36], thus far the researchers have been unable to control the release profile in a specified manner.

In this study we fabricated a method to engineer a novel CNT based delivery vector functionalized with a degradable PLGA coating. Through degradation of PLGA, transcription factors are able to be released in a controlled manner and tune cell behavior. Significant advantages to our proposed system include the ability to transfect cells efficiently with the unique needle-like shape of CNTs, reduce cytotoxicity of pristine CNTs through a biocompatible PLGA coating, and program protein release times by controlling the degradation profiles of the PLGA.

MATERIALS AND METHODS

Carbon Nanotubes Carboxylation

CNTs (0.2g) purchased from Nanolab Inc. with an average length of 1-5 μm, diameter of 15±5 nm, and purity higher than 95% were added into a single neck glass flask with 200 ml 70% nitric acid (Fisher). CNTs were homogenously dispersed by sonication for 60 minutes at ambient conditions. A condenser reflux apparatus was equipped to prevent CNTs from drying out. The reaction temperature was set at 120 °C and after 12 hours the reaction was quenched by addition of 200 ml deionized (DI) water. The mixture was filtered, washed with DI water to neutral pH, and dried at 80 °C overnight[37].

PLGA Functionalization

Carboxyl CNTs (50 mg) were homogeneously dispersed in 50 ml dimethylformamide (DMF) (Fisher) by sonication for 2 hours at ambient conditions. A total of 2 ml oxalyl chloride (Acros) was added drop wise under N2 and stirred in ice water for 2 hours. This was then stirred for 2 hours at room temperature and transferred into an oil bath at 70 °C to be stirred overnight to remove the excess of oxalyl chloride. PLGA (0.5 g) (75:25) (Lactel) was then added to react with carboxylate CNTs and the mixture was stirred at 100 °C for 5 days. Following this, the mixture was cooled down, filtered, and washed with DMF, ethanol (Decon Labs), and DI water. The PLGA linked CNTs (CNT-PLGA) were the black leftover on the filter paper and were dried at 80 °C overnight and collected [38].

Protein Attachment

CNT-PLGA (100 μg) was dispersed in 0.5 ml of 2-(N-Morpholino) ethanosulfonic acid (pH 5.6) (MES) (Acros) buffer under sonication for 1 hour at ambient conditions. 0.25 ml of 1-(3-Dimethylaminopropyl)-3-ethylcarbodiimidehydrochloride (0.2 mol/L) (EDC) (Acros) and 0.25 ml of *N*-Hydroxysuccinimide (0.1 mol/L) (NHS) (Acros) in MES solution were added to the activated carboxylate groups[39,40]. The mixture was washed with PBS and centrifuged in a 100 kDa molecular weight cutoff centrifugal filter (Millipore) to remove EDC and NHS at 5000 g three times for 30 minutes. Then, 5 μg of protein, either bovine serum albumin (BSA) (Sigma), fluorescent BSA (fBSA) (Sigma), or caspase-3 (CP3) (BD) was added into the CNT-PLGA/PBS solution at 4 °C overnight. The mixture was finally washed and centrifuged in a cutoff filter to remove un-conjugated protein six times at 5000 g for 30 minutes[40]. The protein conjugated CNT-PLGA (CNT-PLGA-CP3/BSA/fBSA) solution was collected and stored at -20 °C.

Pro-Ject protein transfection kit (BD) was used as a reference. Liposome nanoparticles were fabricated through the commercial manual. The actual protein dosages were calculated equally to the amount of CNT-PLGA-CP3 groups.

Cell Culture

Osteosarcoma cells MG-63 (ATCC) were cultured in 90% α-MEM (Lonza) and 10% FBS (Gibco) supplemented with 2 mmol/ml l-glutamine (Sigma). Cells were placed in well plate with a seeding density of 20,000/cm2. Testing conjugates were introduced after 4 hours. This time point was set as 0 and cells were then maintained in a humidified incubator at 37°C with 5% CO2.

MTT ASSAY

Cell viability was assessed using 3-(4,5-Dimethylthiazol-2-yl)-2,5-diphenyltetrazoliumbromide (MTT) (Alpha Aesar) calorimetric assay at predetermined time points of 1, 3, 5 and 7 days. In brief, MTT/PBS solution was added into each well (1:5) and incubated at 37 °C for 5 h. This was followed by the removal of medium/MTT solution and addition of 1 ml dimethyl sulfoxide (DMSO) (Fisher). The resulting solution was diluted by DMSO in a ratio of 4:1 and the absorbance was read at 550 nm using Tecan SpectroFluo Plus reader. Cell viability was determined as the equation: ×100%

Spectra measurement

XRD patterns were carried out on a Rigaku MiniFlex II with a scanning speed of 0.2 degrees per minute. UV was measured on a Jasco UV 60 UV-Vis spectrometer from 250 to 500 nm. Fourier Transform Infra-Red (FTIR) spectrum was measured on a Nicolet 6700 FT-IR spectrometer from 3600 to 400 cm-1.

Immunofluorescent images

To remain sterile, cover slips were rinsed in 70% ethanol for 10 minutes and washed with PBS. Cells were seeded onto cover slips and allowed 12 hours to expand on glass slides. The addition of 0.05, 0.1, 0.5, 1 and 3 μg/ml CNT-PLGA-fBSA conjugate was added and incubated for 4 hours. Then samples were washed with PBS, fixed with formalin (Sigma), and dried. Fluorescent images were taken with a Nikon Eclipse 60i microscope system and all images were analyzed with Nikon Elements NX4 software.

TEM

Cells were seeded onto 40 mesh carbon grids for 12 hours with a density of 20,000 cells/cm2followed by an addition of 3 μg/ml CNT-PLGA-BSA conjugate solution and incubated for 4 hours. Samples were then washed with PBS, fixed with formalin, and dehydrated by a series of ethanol treatments[41]. TEM images of the cell penetration of the CNT-PLGA-BSA conjugate were taken with a Hitachi H8000 TEM.

Statistical analysis

Five samples were analyzed at each condition. Data in graphs represents the mean ± standard deviation (SD). Comparison between the two means was determined using the Tukey method and statistical significance was defined as $p \leq 0.05$.

DISCUSSION AND RESULTS

PLGA functionalization

CNTs have been used previously to deliver biological materials across the cell membrane including proteins, DNA, and RNA. Developing the proper surface functionalization for CNTs is the most critical step for a desired application. The two major types of functionalization for CNTs are non-covalent and covalent bonds[42]. Non-covalent binding is based on π-π stacking between the CNTs and aromatic groups from the linkers[43]. Liu et al. showed previously that DNA could be immobilized on pyrene derivatives of functionalized CNTs[44]. Other than these pyrene derivatives, single stranded DNA can also be used to immobilize CNTs[45], however DNA can be cleaved by serum, suggesting non-covalent reactions may not be stable in some cases[46]. For covalent binding, reactive groups are usually formed by various oxidation methods[37,47] which allow for further modifications to enhance polymer[38], protein[48] and DNA[33] attachment. Functionalized CNTs can also be easily dispersed in water, which allows for forming supermolecular bioconjugates such as PLGA and PEG. PLGA is a FDA approved polymer for clinic use with the unique ability to control degradation rate, which allows us to tune the drug release profile. PEG is a commercially manufactured clinical material which has been previously conjugated to CNTs in order to deliver siRNA through the instant cleavage of the S-S bond. PEG has a single functional group and lower average molecular weight when compared to PLGA. Further modifications are thus required to attach active biological molecules to PEG and control the release profile. To this end, a novel CNT-PLGA based drug delivery system has several advantages including a high transfection rate, reduced toxicity, and a highly controlled drug release profile.

In this study, a novel CNT based drug delivery system as depicted in Figure 1 was developed. In general, pristine CNTs as shown in Figure 2 were oxidized in nitric acid to form carboxyl groups. Carboxyl groups of CNTs were then activated by oxalyl chloride and attached to the PLGA[38]. PLGA is able to provide attachment for the desired transcription factor

CP3. The CNT-PLGA-CP3 conjugate can be easily dispersed in PBS without aggregation, which otherwise would hinder penetration of the cell membrane. Due to the enzymatic degradation of PLGA, CP3 is able to be released gradually and induce cell apoptosis. The CNT-PLGA-CP3 conjugations are stable for weeks at -20 °C and the delivery profile can be tuned simply by changing the concentrations of PLGA.

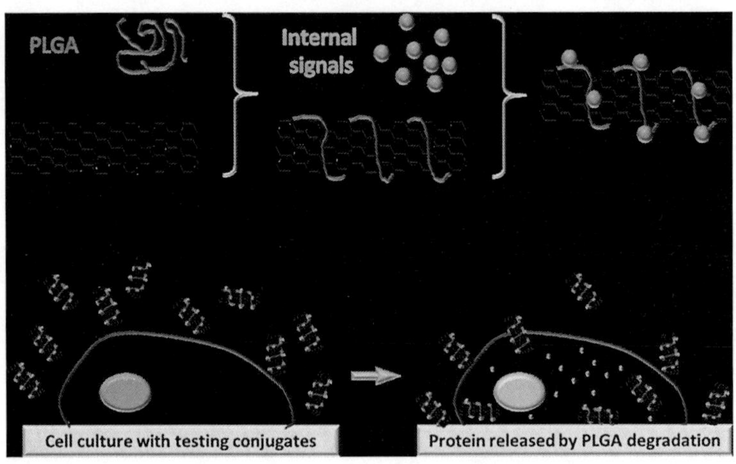

Figure 1. Schematic representation of CNT-PLGA conjugate fabrication and intercellular delivery

CNTs are carboxylated, coated with PLGA, and functionalized with caspase-3. The conjugates are able to penetrate into MG-63 cells and release caspase-3 through the degradation of PLGA.

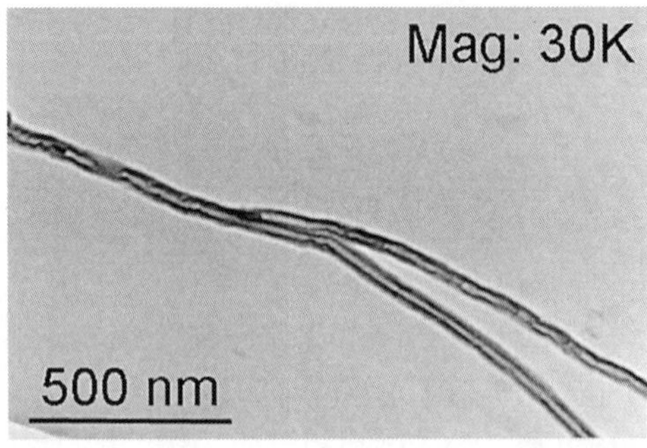

Figure 2. Pristine carbon nanotubes prior to conjugation.

TEM image of CNTs. The results show that CNTs have no outer coating prior to PLGA and protein conjugation.

XRD, UV, IR and TEM were used to investigate the CNT-PLGA complex. In Figure 3A, all XRD patterns were roughly the same in the region larger than 35 degrees which represents the crystal structure of the C-H bond, *sp*2 hybrid C-C bond, and C-O bond [49]. However, PLGA patterns showed a peak, which denoted the amorphous region of PLGA polymers from 10-25 degrees as shown in Figure 3B. This peak can be determined as a characteristic of PLGA, which CNTs alone do not possess. A small peak at 22.7 degrees for amorphous PLGA and a peak at 24.1 degrees for CNTs were observed, indicating PLGA attachment to CNTs. However, this peak was not as pronounced as observed in pattern A. We believe only small amounts of PLGA were able to bind to CNTs resulting in a decreased signal. CNTs have a simple flat UV absorption and start to decrease from 280 nm as shown in Figure 3C. PLGA didn't show a similar trend, instead it showed a strong absorption at 260 nm similar to other studies[50]. CNT-PLGA showed absorption at 260 nm indicating PLGA and a flat decreasing trend thereafter, indicating CNTs were present. This finding showed a similar trend to CNT-Paclitaxel[34]. Through FTIR spectra CNTs showed a typical *sp*2 hybrid C-C stretching at 1200 cm-1, while no O-H stretching was observed at 3400 cm-1 as shown in Figure 4A. On the contrary, O-H stretching and C=O stretching at 1650 cm-1 were observed in CNT-PLGA. This denoted a successful attachment of PLGA to the CNTs. In addition to FTIR, TEM was used to observe the PLGA linkage to CNTs. Pristine multi-wall CNTs in Figure 2 with no PLGA functionalization was compared to the dark PLGA layer clearly shown coated over the CNTs with a depth of several nanometers in Figure 3D.

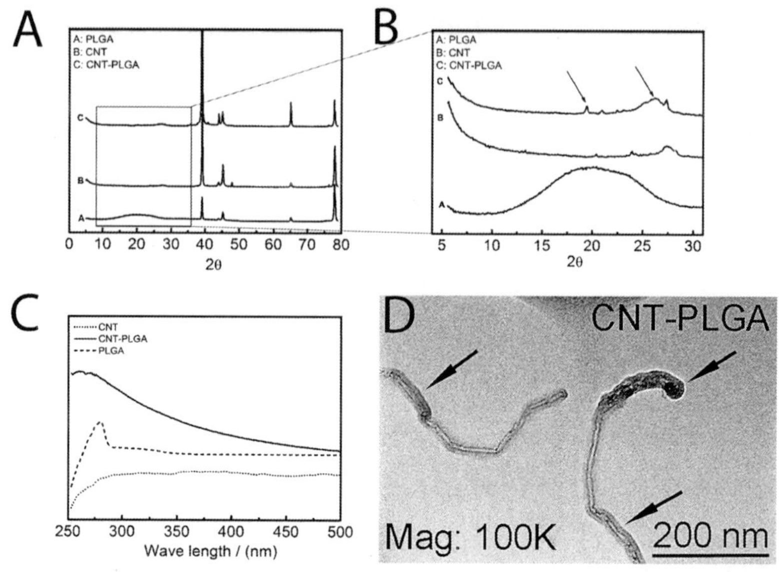

Figure 3. Characterization of CNT-PLGA complex.

(A) XRD pattern (B) XRD pattern (C) UV-Vis spectrum (D) TEM image. Results were able to show that PLGA is successfully conjugated to CNTs.

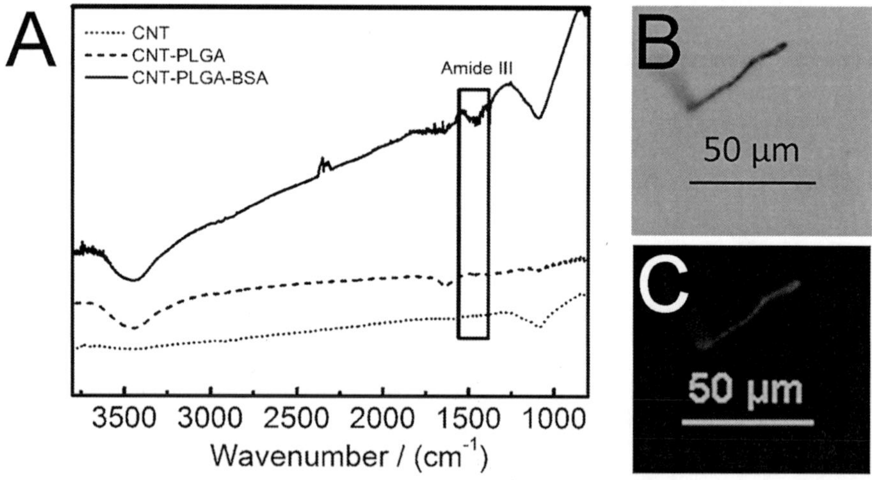

Figure 4. Protein linkage to CNT-PLGA conjugates.

(A) Infra-red spectrum (B) light microscopy of CNT-PLGA-BSA (C) fluorescent images of CNT-PLGA-fBSA. Results show that protein has been successfully attached to CNT-PLGA complex, forming CNT-PLGA-BSA conjugates.

Attachment

DNA and RNA are the genetic material generally transported by functionalized CNTs. In 2004, Pantarotto discovered CNTs as a tool to transport DNA plasmids[25]. Further research found DNA was protected by CNTs during cellular uptake[51]. In addition, Zhang et al. successfully transported telomerase inhibited small interference RNA into tumor cells and suppressed their growth[52]. However, researchers generally used protein fictionalization for stabilizing [53], labeling[54] and separating[55] CNTs rather than transporting functional protein to tune cellular behavior. Here, we were able to conjugate protein to the CNT-PLGA complex.

Protein typically has a UV absorption at 280 nm due to *sp*2 hybrid C-C in Phe, Tyr, and Trp which is a similar structure to CNTs. Protein also has a similar amorphous region in XRD pattern as most polymers. Therefore, XRD and UV are not appropriate methods to confirm protein attachment so we used FTIR and fluorescently tagged BSA to confirm the attachment of protein. As shown in Figure 4A, CNT-PLGA-BSA showed a peak around 1400 cm-1indicating amide III belt, which is the typical peak of BSA. fBSA was then used to attach to CNT-PLGA and in Figure 4B and 4C, a single CNT-PLGA attached to fBSA was observed through immunofluorescence indicating successful conjugation. CNTs are generally auto fluorescent over 800 nm [56] but no fluorescent signal was observed at those wavelengths confirming the protein attachment.

Cellular Transfection by CNT-PLGA-fBSA conjugates

The ideal vector to deliver biological material should have two basic characteristics being high efficiency and safety. In terms of efficiency, we investigated the cellular uptake ability of CNT-PLGA-fBSA and CNTs were found to have the ability to penetrate cells due to their nano-sized diameter and small radius to volume ratio[57]. In demonstrating the high efficiency to penetrate cells, we tested the transfection ability of CNT-PLGA-fBSA conjugations in different dosages. In order to prevent non-specific background signals, cells were carefully washed to remove any non-penetrated CNT-PLGA-fBSA conjugates. We were able to observe fluorescent signals in cells as shown in Figure 5. The CNT-PLGA-fBSA conjugates showed a pronounced ability to penetrate cells with a transfection rate close to 100% at all conducted concentrations, ranging from 0.05 to 3 μg/ml. The high magnification image of Figure 6A, 6B, and 6C are able to show higher resolutions around the nuclei which shows higher fluorescent signal, coinciding with the location of where large amounts of ribosomes reside. This positioning of the CNT-PLGA-fBSA conjugates facilitates RNA and transcription factor delivery to the cells. TEM was also used to confirm cell penetration as shown in Figure 6D, where clearly there is a CNT-PLGA-BSA conjugate penetrating into the cytosol, demonstrating the ability of CNTs to transport material across the membrane. We have successfully achieved a transfection rate higher than the 30% or 40% attained by polymers and liposome nanoparticles indicating the promise in non-viral gene delivery vectors.

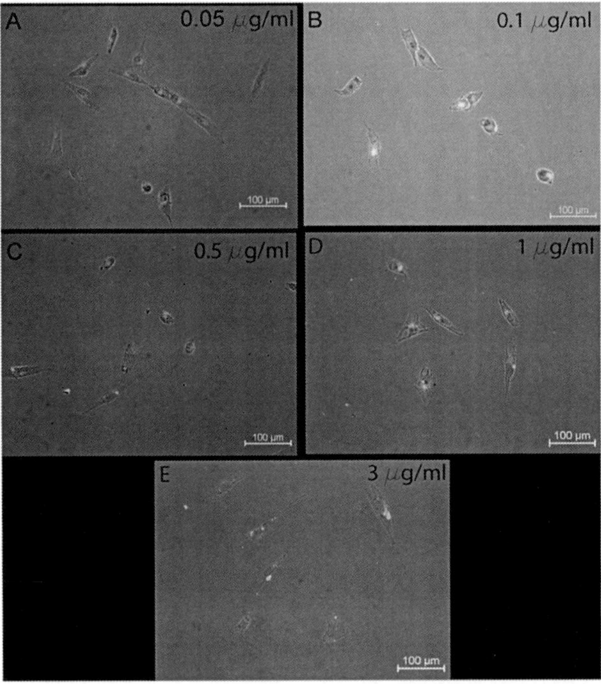

Figure 5. Optic and fluorescent combined images of CNT-PLGA-fBSA delivery into MG-63 osteosarcoma cells.

The dosages with osteosarcoma cells were 0.05 μg/ml, 0.1 μg/ml, 0.5 μg/ml, and 1 μg/ml and 3 μg/ml, respectively. The results showed that CNT-PLGA-fBSA could easily penetrate into cells and transduce osteosarcoma cells. All images are 20x magnification.

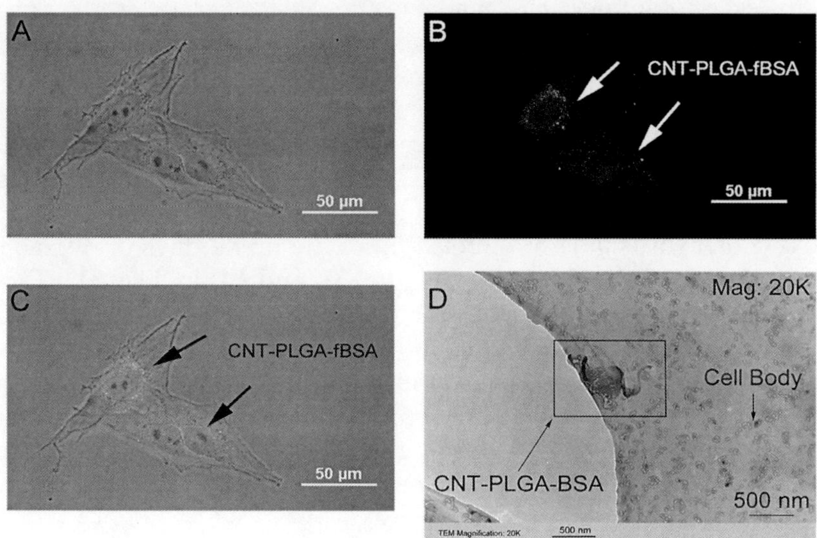

Figure 6. Penetration of CNT-PLGA-proteins into the cells.

Determination of CNTs delivered into osteosarcoma cells. Dosage is 3 μg/ml. (A) Osteosarcoma cells cultured with CNT-PLGA-fBSA (B) Image of fluorescent signal within osteosarcoma cells (C) Optic and fluorescent combined image of CNT-PLGA-fBSA image (D) TEM image of a single CNT-PLGA-BSA penetrating the cell body.

The efficacy of CNT-PLGA-CP3 in tuning apoptosis

Commonly used drugs for cancer therapies are alkylating agents, anti-metabolites, plant alkaloid and terpenoids, topoisomerase inhibitors, and cytotoxic antibiotics. These chemical agents generally present side effects which not only kill cancer cells, but also induce necrosis of healthy tissue. In order to prevent or limit these side effects, we have chosen CP3 as a potential candidate. CP3 is an enzyme that is highly involved in the cell apoptosis pathway. We used MTT assays to test cell viability as a standard to measure the ability of CNT-PLGA-CP3 to release CP3 to cells. The working concentration is set at low levels in order to eliminate cell necrosis induced by large amounts of CNTs. Researchers have demonstrated functionalized CNTs were found to be less toxic than pristine CNTs[58,59], and pristine CNTs in low concentrations have been shown to exhibit acceptable toxicity levels[60]. So cell death experienced in this experiment is attributed to the contribution of CP3 delivered to cells inducing apoptosis.

Pro-Ject is a commercially available liposome generally used for protein delivery which we used as a positive control in comparison to our CNT-PLGA-CP3 conjugation. As shown in Figure 7A, no significant differences were observed in CNT-PLGA-CP3 groups on day 1, however, Pro-Ject groups showed significant difference in comparison. We attribute this to the fact that the degradation of PLGA hasn't fully allowed the release of significant amounts of CP3[61]. We went on to further demonstrate that limited protein was able to be released in the first 2 days of degradation with an *in vitro* release profile (Figure S1 and S2 in File S1). Pro-Ject liposomes released material with minimal differences in dosage on Day 1. We hypothesize that the average amount of CP3 delivered was significant enough to induce apoptosis even at minimal concentrations. Interestingly, the cell viability of all Pro-Ject liposome samples average at approximately 50%. Considering the CP3 was enough to induce apoptosis, the efficiency of Pro-Ject liposome groups was close to 50%.

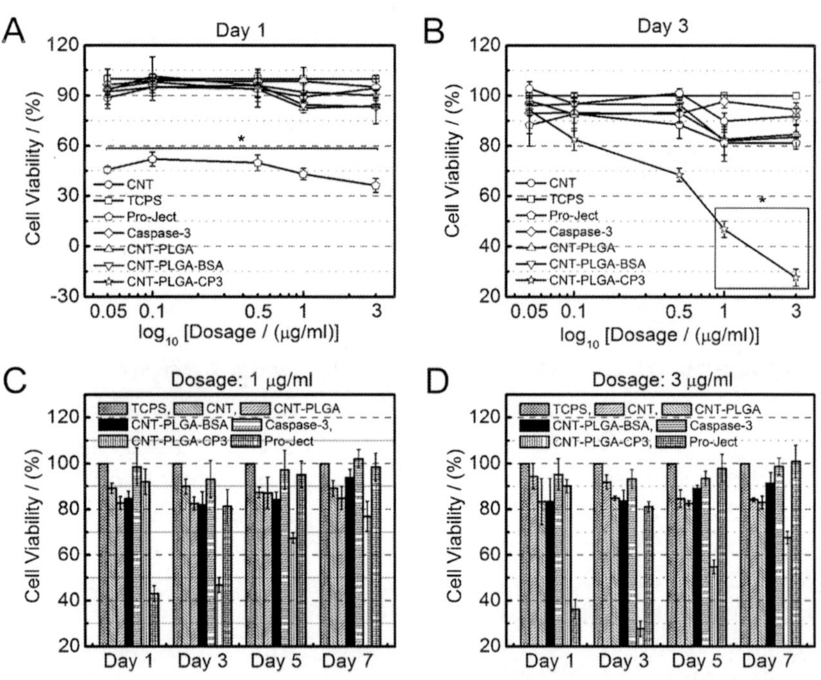

Figure 7. Carbon nanotubes have the ability to penetrate cells and release caspase-3.

Cell viability on day 1 (A) and day 3 (B) of MG-63 cells under exposure to differing dosages of CNTs, CNT-PLGA, CNT-PLGA-BSA, caspase-3,

Pro-Ject and CNT-PLGA-CP3 treatment. * means significant difference of cell viability under CNT-PLGA-CP3 cojugate exposure and other groups of same dosage.

Figure 7B shows the cell viability of CNT-PLGA-CP3 treated samples were significantly smaller compared to positive and negative controls on Day 3. Pro-Ject liposome groups treated samples showed minimal differences in this case, indicating there was little to no consistency of material delivery over long periods. Conjugate dosage played an important role in inducing cell apoptosis with no significant differences observed between CNT-PLGA-CP3 in low dosages and controls while CNT-PLGA-CP3 conjugates in high concentration exhibited a significantly low cell viability compared to other controls. We believe the amount of transcription factor reacting with the CNT-PLGA complex is the same but the amount of CP3 released was highly dependent on the conjugate dosage.

We also conducted long-term cell viability tests at two high dosages with no additional treatments throughout the experiment. The results shown in Figure 7C and 7D show that cell viability was still suppressed by treating 1 µg/ml CNT-PLGA-CP3 on Day 5 and 3 µg/ml CNT-PLGA-CP3 on Day 7, respectively. Note that cancer cells are able to recover and become fully confluent under Pro-Ject liposome groups between Day 1 and Day 3. We believe CP3 is superior due to the gradual release profile.

CONCLUSIONS

We were able to successfully fabricate a CNT-PLGA system, which is able to deliver biological material including genes, transcriptional factors, and signal molecules into cells. This system has shown a high transfection rate and a reliable, time dependent release profile. Yet another advantage to this system is that the releasing profile can be tuned simply by controlling the molecular weight and ratio of PLGA (Figures S3 and S4 in File S1). In all aspects, this CNT-PLGA-CP3 conjugation is a highly efficient and promising drug delivery system with possible future applications in developing scaffolds for bone tissue engineering.

AUTHOR CONTRIBUTIONS

Conceived and designed the experiments: QC EJ. Performed the experiments: QC MB. Analyzed the data: QC EJ. Contributed reagents/materials/analysis tools: QC MB. Wrote the manuscript: QC GH EJ.

REFERENCES

1. Stewart BW, Kleihues P, International Agency for Research on Cancer (2003) World cancer report. Lyon: IARC Press. 351 pp.
2. Blaese RM (1990) The Ada Human Gene-Therapy Clinical Protocol. Hum Gene Ther 1: 327-329. doi:10.1089/hum.1990.1.3-327.
3. Sun YS, Finger C, Alvarez-Vallina L, Cichutek K, Buchholz CJ (2005) Chronic gene delivery of interferon-inducible protein 10 through replication-competent retrovirus vectors suppresses tumor growth. Cancer Gene Ther 12: 900-912. doi:10.1038/sj.cgt.7700854.
4. Hoffman RM (2007) Noninvasive imaging for evaluation of the systemic delivery of capsid-modified adenovirus in an orthotopic model of advanced lung cancer. Cancer 109: 1213-1213. doi:10.1002/cncr.22494.
5. Marks WJ, Ostrem JL, Verhagen L, Starr PA, Larson PS et al. (2008) Safety and tolerability of intraputaminal delivery of CERE-120 (adeno-associated virus serotype 2-neurturin) to patients with idiopathic Parkinson's disease: an open-label, phase I trial. Lancet Neurol 7: 400-408. doi:10.1016/S1474-4422(08)70065-6.
6. Anderson WF (1998) Human gene therapy. Nature 392: 25-30. doi:10.1038/32058.
7. Matsumura Y, Maeda H (1986) A New Concept for Macromolecular Therapeutics in Cancer-Chemotherapy - Mechanism of Tumoritropic Accumulation of Proteins and the Antitumor Agent Smancs. Cancer Res 46: 6387-6392.
8. Peer D, Karp JM, Hong S, FaroKhzad OC, Margalit R et al. (2007) Nanocarriers as an emerging platform for cancer therapy. Nat Nanotechnol 2: 751-760. doi:10.1038/nnano.2007.387.
9. Yang XZ, Dou S, Sun TM, Mao CQ, Wang HX et al. (2011) Systemic delivery of siRNA with cationic lipid assisted PEG-PLA nanoparticles for cancer therapy. J Control Release 156: 203-211. doi:10.1016/j.jconrel.2011.07.035.
10. Na K, Lee KH, Lee DH, Bae YH (2006) Biodegradable thermo-sensitive nanoparticles from poly(L-lactic acid)/poly(ethylene glycol) alternating multi-block copolymer for potential anti-cancer drug carrier. Eur J Pharm Sci 27: 115-122. doi:10.1016/j.ejps.2005.08.012.
11. Xu JS, Huang JW, Qin RG, Hinkle GH, Povoski SP et al. (2010) Synthesizing and binding dual-mode poly (lactic-co-glycolic acid) (PLGA) nanobubbles for cancer targeting and imaging. Biomaterials 31: 1716-1722. doi:10.1016/j.biomaterials.2009.11.052.
12. Kim TH, Jin H, Kim HW, Cho MH, Cho CS (2006) Mannosylated chitosan nanoparticle-based cytokine gene therapy suppressed cancer growth in BALB/c mice bearing CT-26 carcinoma cells. Mol Cancer Ther 5: 1723-1732. doi:10.1158/1535-7163.MCT-05-0540.
13. Nicklas M, Schatton W, Heinemann S, Hanke T, Kreuter J (2009) Preparation and characterization of marine sponge collagen nanoparticles and

employment for the transdermal delivery of 17 beta-estradiol-hemihydrate. Drug Dev Ind Pharm 35: 1035-1042. doi:10.1080/03639040902755213.

14. Ge J, Min SH, Kim DM, Lee DC, Park KC et al. (2012) Selective gene delivery to cancer cells secreting matrix metalloproteinases using a gelatin/polyethylenimine/DNA complex. Biotechnology and Bioprocess Engineering 17: 160-167. doi:10.1007/s12257-011-0423-x.

15. Gul-Uludag H, Xu P, Marquez-Curtis LA, Xing J, Janowska-Wieczorek A et al. (2012) Cationic liposome-mediated CXCR4 gene delivery into hematopoietic stem/progenitor cells: implications for clinical transplantation and gene therapy. Stem Cells Dev 21: 1587-1596. doi:10.1089/scd.2011.0297.

16. Baughman RH, Zakhidov AA, de Heer WA (2002) Carbon nanotubes - the route toward applications. Science 297: 787-792. doi:10.1126/science.1060928.

17. Kim SK, Day RW, Cahoon JF, Kempa TJ, Song KD et al. (2012) Tuning Light Absorption in Core/Shell Silicon Nanowire Photovoltaic Devices through Morphological. Design - Nano Lett 12: 4971-4976. doi:10.1021/nl302578z.

18. Goldberger J, He R, Zhang Y, Lee S, Yan H et al. (2003) Single-crystal gallium nitride nanotubes. Nature 422: 599-602. doi:10.1038/nature01551.

19. Chopra NG, Luyken RJ, Cherrey K, Crespi VH, Cohen ML et al. (1995) Boron-Nitride Nanotubes. Science 269: 966-967. doi:10.1126/science.269.5226.966.

20. Mor GK, Varghese OK, Paulose M, Shankar K, Grimes CA (2006) A review on highly ordered, vertically oriented TiO2 nanotube arrays: Fabrication, material properties, and solar energy applications. Solar Energy Materials and Solar Cells 90: 2011-2075. doi:10.1016/j.solmat.2006.04.007.

21. Wang ZL (2004) Zinc oxide nanostructures: growth, properties and applications. Journal of Physics:_Condensed Matter 16: R829-R858. doi:10.1088/0953-8984/16/25/R01.

22. Iijima S, Ichihashi T (1993) Single-Shell Carbon Nanotubes of 1-Nm Diameter. Nature 363: 603-605. doi:10.1038/363603a0.

23. Hilder TA, Hill JM (2008) Probability of encapsulation of paclitaxel and doxorubicin into carbon nanotubes. Microbiol-- Nano Letters 3: 41-49.

24. Weng XX, Wang MY, Ge J, Yu SN, Liu BH et al. (2009) Carbon nanotubes as a protein toxin transporter for selective HER2-positive breast cancer cell destruction. Mol Biosyst 5: 1224-1231. doi:10.1039/b906948h.

25. Pantarotto D, Singh R, McCarthy D, Erhardt M, Briand JP et al. (2004) Functionalized carbon nanotubes for plasmid DNA gene delivery. Angew Chem Int Ed Engl 43: 5242-5246. doi:10.1002/anie.200460437.

26. Singh R, Pantarotto D, McCarthy D, Chaloin O, Hoebeke J et al. (2005) Binding and condensation of plasmid DNA onto functionalized carbon nanotubes: toward the construction of nanotube-based gene delivery vectors. J Am Chem Soc 127: 4388-4396. doi:10.1021/ja0441561.

27. Ladeira MS, Andrade VA, Gomes ERM, Aguiar CJ, Moraes ER et al. (2010) Highly efficient siRNA delivery system into human and murine cells using single-wall carbon nanotubes. Nanotechnology 21: 385101.

28. Jin H, Heller DA, Strano MS (2008) Single-particle tracking of endocytosis and exocytosis of single-walled carbon nanotubes in NIH-3T3 cells. Nano Lett 8: 1577-1585. doi:10.1021/nl072969s.

29. Wang MY, Yu SN, Wang CA, Kong JL (2010) Tracking the Endocytic Pathway of Recombinant Protein Toxin Delivered by Multiwalled Carbon Nanotubes. Acs Nano 4: 6483-6490. doi:10.1021/nn101445y.

30. Donaldson K, Stone V, Tran CL, Kreyling W, Borm PJA (2004). Nanotoxicology - Occupational and Environmental Medicine 61: 727-728. doi:10.1136/oem.2004.013243.

31. Mikael PE, Nukavarapu SP (2011) Functionalized Carbon Nanotube Composite Scaffolds for Bone Tissue Engineering: Prospects and Progress. Journal Biomaterials and Tissue Engineering 1: 76-85. doi:10.1166/jbt.2011.1011.

32. Singh R, Pantarotto D, Lacerda L, Pastorin G, Klumpp C et al. (2006) Tissue biodistribution and blood clearance rates of intravenously administered carbon nanotube radiotracers. Proc Natl Acad Sci U S A 103: 3357-3362.

33. Liu Z, Fan AC, Rakhra K, Sherlock S, Goodwin A et al. (2009) Supramolecular stacking of doxorubicin on carbon nanotubes for in vivo cancer therapy. Angew Chem Int Ed Engl 48: 7668-7672. doi:10.1002/anie.200902612.

34. Liu Z, Chen K, Davis C, Sherlock S, Cao Q et al. (2008) Drug delivery with carbon nanotubes for in vivo cancer treatment. Cancer Res 68: 6652-6660. doi:10.1158/0008-5472.CAN-08-1468.

35. Liu Z, Winters M, Holodniy M, Dai H (2007) siRNA delivery into human T cells and primary cells with carbon-nanotube transporters. Angew Chem Int Ed Engl 46: 2023-2027. doi:10.1002/anie.200604295.

36. Heller DA, Baik S, Eurell TE, Strano MS (2005) Single-walled carbon nanotube spectroscopy in live cells: Towards long-term labels and optical sensors. Advanced Materials 17: 2793- +.

37. Rosca ID, Watari F, Uo M, Akaska T (2005) Oxidation of multiwalled carbon nanotubes by nitric acid. Carbon 43: 3124-3131. doi:10.1016/j.carbon.2005.06.019.

38. Zhao B, Hu H, Yu AP, Perea D, Haddon RC (2005) Synthesis and characterization of water soluble single-walled carbon nanotube graft copolymers. J Am Chem Soc 127: 8197-8203. doi:10.1021/ja042924i.

39. McCarron PA, Marouf WM, Donnelly RF, Scott C (2008) Enhanced surface attachment of protein-type targeting ligands to poly(lactide-co-glycolide) nanoparticles using variable expression of polymeric acid functionality. J Biomed Mater Res A 87A: 873-884. doi:10.1002/jbm.a.31835.

40. Liu Z, Tabakman SM, Chen Z, Dai HJ (2009) Preparation of carbon nanotube bioconjugates for biomedical applications. Nat Protoc 4: 1372-1382. doi:10.1038/nprot.2009.146.

41. Cheng Q, Rutledge K, Jabbarzadeh E (2013) Carbon Nanotube–Poly(lactide-co-glycolide) Composite Scaffolds for Bone Tissue Engineering Applications. Annals of Biomedical Engineering.

42. Liu Z, Tabakman S, Welsher K, Dai HJ (2009) Carbon Nanotubes in Biology and Medicine: In vitro and in vivo Detection, Imaging and Drug Delivery. Nano Research 2: 85-120. doi:10.1007/s12274-009-9009-8.

43. Chen J, Liu HY, Weimer WA, Halls MD, Waldeck DH et al. (2002) Noncovalent engineering of carbon nanotube surfaces by rigid, functional conjugated polymers. J Am Chem Soc 124: 9034-9035. doi:10.1021/ja026104m.

44. Liu Y, Yu ZL, Zhang YM, Guo DS, Liu YP (2008) Supramolecular architectures of beta-cyclodextrin-modified chitosan and pyrene derivatives mediated by carbon nanotubes and their DNA condensation. J Am Chem Soc 130: 10431-10439. doi:10.1021/ja802465g.

45. Zheng M, Jagota A, Semke ED, Diner BA, Mclean RS et al. (2003) DNA-assisted dispersion and separation of carbon nanotubes. Nat Mater 2: 338-342. doi:10.1038/nmat877.

46. Moon HK, Il Chang C, Lee DK, Choi HC (2008) Effect of Nucleases on the Cellular Internalization of Fluorescent Labeled DNA-Functionalized Single-Walled Carbon Nanotubes. Nano Research 1: 351-360. doi:10.1007/s12274-008-8038-z.

47. Niyogi S, Hamon MA, Hu H, Zhao B, Bhowmik P et al. (2002) Chemistry of single-walled carbon nanotubes. Acc Chem Res 35: 1105-1113. doi:10.1021/ar010155r.

48. Chen Z, Tabakman SM, Goodwin AP, Kattah MG, Daranciang D et al. (2008) Protein microarrays with carbon nanotubes as multicolor Raman labels. Nat Biotechnol 26: 1285-1292. doi:10.1038/nbt.1501.

49. O'Connell M (2006) Carbon nanotubes : properties and applications. Boca Raton, FL: CRC/Taylor & Francis. 319 p. p..

50. Cheng FY, Su CH, Wu PC, Yeh CS (2010) Multifunctional polymeric nanoparticles for combined chemotherapeutic and near-infrared photothermal cancer therapy in vitro and in vivo. Chem Commun (Camb) 46: 3167-3169. doi:10.1039/b919172k. PubMed:20424762.

51. Wu Y, Phillips JA, Liu H, Yang R, Tan W (2008) Carbon nanotubes protect DNA strands during cellular delivery. ACS Nano 2: 2023-2028. doi:10.1021/nn800325a.

52. Zhang ZH, Yang XY, Zhang Y, Zeng B, Wang ZJ et al. (2006) Delivery of telomerase reverse transcriptase small interfering RNA in complex with positively charged single-walled carbon nanotubes suppresses tumor growth. Clinical Cancer Research 12: 4933-4939. doi:10.1158/1078-0432.CCR-05-2831..

53. Karajanagi SS, Yang HC, Asuri P, Sellitto E, Dordick JS et al. (2006) Protein-assisted solubilization of single-walled carbon nanotubes. Langmuir 22: 1392-1395. doi:10.1021/la0528201.

54. Yoshimura SH, Khan S, Maruyama H, Nakayama Y, Takeyasu K (2011) Fluorescence Labeling of Carbon Nanotubes and Visualization of a Nanotube-Protein Hybrid under Fluorescence Microscope. Biomacromolecules 12: 1200-1204. doi:10.1021/bm101491s.

55. Diao XH, Chen HY, Zhang GL, Zhang FB, Fan XB (2012) Magnetic Carbon Nanotubes for Protein Separation. Journal of Nanomaterials.

56. Ebbesen TW (1997) Carbon nanotubes : preparation and properties. Boca Raton: CRC Press. 296 pp.

57. Lacerda L, Raffa S, Prato M, Bianco A, Kostarelos K (2007) Cell-penetrating CNTs for delivery of therapeutics. Nano Today 2: 38-43. doi:10.1016/S1748-0132(07)70172-X.

58. Chen X, Tam UC, Czlapinski JL, Lee GS, Rabuka D et al. (2006) Interfacing carbon nanotubes with living cells. J Am Chem Soc 128: 6292-6293. doi:10.1021/ja060276s.

59. Sayes CM, Wahi R, Kurian PA, Liu Y, West JL et al. (2006) Correlating nanoscale titania structure with toxicity: a cytotoxicity and inflammatory response study with human dermal fibroblasts and human lung epithelial cells. Toxicol Sci 92: 174-185. doi:10.1093/toxsci/kfj197.

60. Lacerda L, Bianco A, Prato M, Kostarelos K (2006) Carbon nanotubes as nanomedicines: from toxicology to pharmacology. Adv Drug Deliv Rev 58: 1460-1470. doi:10.1016/j.addr.2006.09.015.

61. Anderson JM, Shive MS (1997) Biodegradation and biocompatibility of PLA and PLGA microspheres. Advanced Drug Delivery Reviews 28: 5-24. doi:10.1016/S0169-409X(97)00048-3.

Chapter 5

PET IMAGING OF SOLUBLE YTTRIUM-86-LABELED CARBON NANOTUBES IN MICE

Michael R. Mcdevitt, Debjit Chattopadhyay, Jaspreet S. Jaggi, Ronald D. Finn, Pat B. Zanzonico, Carlos Villa, Diego Rey, Juana Mendenhall, Carl A. Batt, Jon T. Njardarson, And David A. Scheinberg

Molecular Pharmacology and Chemistry Department, Departments of Medicine, Radiology, and Medical Physics, Memorial Sloan-Kettering Cancer Center, New York, New York, United States of America

Department of Biomedical Engineering, Cornell University, Ithaca, New York, United States of America

Department of Food Science, Cornell University, Ithaca, New York, United States of America

Department of Chemistry and Chemical Biology, Cornell University, Ithaca, New York, United States of America

ABSTRACT

The potential medical applications of nanomaterials are shaping the landscape of the nanobiotechnology field and driving it forward.

A key factor in determining the suitability of these nanomaterials must be how they interface with biological systems. Single walled carbon nanotubes (CNT) are being investigated as platforms for the delivery of biological, radiological, and chemical payloads to target tissues. CNT are mechanically robust graphene cylinders comprised of sp^2-bonded carbon atoms and possessing highly regular structures with defined periodicity. CNT exhibit unique mechano chemical properties that can be exploited for the development of novel drug delivery platforms. In order to evaluate the potential usefulness of this CNT scaffold, we undertook an imaging study to determine the tissue bio distribution and pharmacokinetics of prototypical DOTA-functionalized CNT labeled with yttrium-86 and indium-111 (^{86}Y-CNT and ^{111}In-CNT, respectively) in a mouse model.

Methodology and Principal Findings

The ^{86}Y-CNT construct was synthesized from amine-functionalized, water-soluble CNT by covalently attaching multiple copies of DOTA chelates and then radiolabeling with the positron-emitting metal-ion, yttrium-86. A gamma-emitting ^{111}In-CNT construct was similarly prepared and purified. The constructs were characterized spectroscopically, microscopically, and chromatographically. The whole-body distribution and clearance of yttrium-86 was characterized at 3 and 24 hours post-injection using positron emission tomography (PET). The yttrium-86 cleared the blood within 3 hours and distributed predominantly to the kidneys, liver, spleen and bone. Although the activity that accumulated in the kidney cleared with time, the whole-body clearance was slow. Differential uptake in these target tissues was observed following intraveneous or intraperitoneal injection.

The whole-body PET images indicated that the major sites of accumulation of activity resulting from the administration of ^{86}Y-CNT were the kidney, liver, spleen, and to a much less extent the bone. Blood clearance was rapid and could be beneficial in the use of short-lived radionuclides in diagnostic applications.

INTRODUCTION

How nanomaterials interface with biological systems is a key question that may impact the emerging field of nanomedicine. The pharmacokinetic profile of these unique, new materials is a prominent factor in determining their suitability for *in vivo* applications. The important issues of where they distribute *in vivo* and how they clear from a living system must be addressed. Single walled carbon nanotubes [1] (CNT) are promising scaffolds for transporting biological cargo across cellular membranes [2]–[5]. A report [6] of the murine biodistribution of an ^{125}I-labeled hydroxylated-CNT into the stomach, kidney, and bone was contrasted by another report[7] describing the rapid blood and whole body clearance of an ^{111}In-labeled diethylenetriaminepentaacetic acid (DTPA) derivatized-CNT from mice. The pharmacokinetics of an unmodified-CNT suspended in surfactant was also reported [8] in rabbits utilizing the inherent CNT near-infrared fluorescence for detection and demonstrated rapid blood clearance and only liver accumulation. We attached 1,4,7,10-tetraazacyclododecane-1,4,7,10-tetraaceticacid (DOTA), a radiometal chelate moiety, to soluble, amino-functionalized-CNT. [9],[10] and radiolabeled with the short half-lived, positron-emitting ^{86}Y ($t_{1/2}$ = 14.7 h) or the long half-lived, gamma-emitting ^{111}In ($t_{1/2}$ = 2.81 d). The goal of this study was to map the *in vivo*distribution of our soluble prototype CNT scaffold in real time and evaluate the pharmacokinetic profile using PET and conventional tissue harvest techniques. Herein, we report the PET imaging and biodistribution of a positron-emitting CNT construct in mice. PET was chosen because it provides extremely sensitive, quantitative, and functional information that is different from that obtainable with other largely anatomical imaging modalities and CT was performed to confirm anatomical assignments [11].

METHODS

Synthesis and Characterization of ^{86}Y-CNT

Pristine single walled carbon nanotubes (CNT) were shortened and purified by oxidative acid digestion. Briefly, 60 mg of CNT (Nanostructured & Amorphous Materials, Los Alamos, NM, Lot # 1280YJ) were refluxed in 2M HNO_3 (60 mL) for 48 hours. In addition, sonication was performed at 70°C with 100 watts for 0.5 h with a

Biologics Model 300 V/T Ultrasonic Homogenizer (Manassas, VA) using the 3/8" diameter stepped titanium tip. The CNT were again refluxed for an additional 10 h in the 2M HNO_3. This digestion and sonication yields purified and shortened carboxy-functionalized CNT molecules (Figure 1, Compound **1**). Raman spectroscopy was employed to characterize **1** using a Renishaw InVia microRaman system (Gloucestershire, United Kingdom) equipped with a 488 nm laser. Samples were evaporated on silicon substrates.

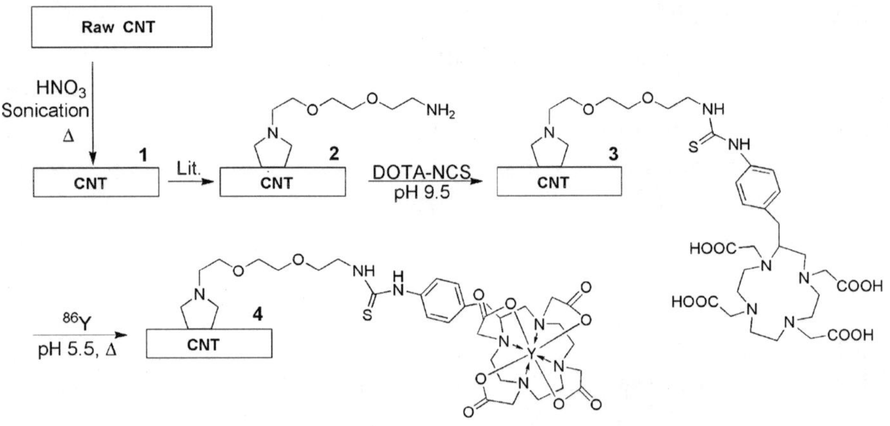

Figure 1. Synthetic scheme to prepare ^{86}Y-CNT.

The carboxy-functionalized CNT molecules were functionalized and subsequently solublized by covalent attachment of reactive sidewall amino groups [9], [10] yielding CNT-(NH_2) (Figure 1, compound **2**). Raman spectroscopy was employed to characterize **2** (as described above) and the number of primary amines that were appended per mass of CNT was measured using the Sarin assay [12].

The water soluble CNT-amine construct, **2**, was reacted with 2-(p-isothiocyanatobenzyl)-1,4,7,10-tetraazacyclododecane-1,4,7,10-tetraaceticacid (DOTA-NCS, Macrocyclics, Inc., Dallas, TX) to yield a CNT-DOTA construct in metal-free conditions at pH 9.5 (adjusted with 1 M metal free carbonate solution) for 40 minutes at room temperature at a stoichiometry of 2:1 (DOTA-NCS to amine). The product was purified using a 10 DG gel permeation column (BioRad, Hercules, CA) with metal-free water (MFW, Purelab Plus System, U.S. Filter Corp., Lowell, MA) as the mobile phase. The 10 DG column was rendered metal free by washing with 30 mL of 10 mM EDTA followed by rinsing with 60 mL of MFW. The product was lyophilized to yield a solid that was the CNT-DOTA construct (Figure 1, compound **3**). Raman spectroscopy was employed to characterize **3** (as described above). AFM images were obtained using a Veeco Dimension 3100 (Woodbury, NY) operated in the tapping mode with a scanning frequency of 84 Hz and a scan rate of 0.500 Hz. Cantilever tips were supplied from Vecco probes (force constant of 3 k/m). Samples were spin coated onto a freshly cleaved mica surface. TEM images were obtained using a Philips EM-201 (Amsterdam, Netherlands) with 1 nm resolution, HV = 80 kV. Samples are adsorbed onto plasma-treated formvar-coated copper grids. Ultraviolet-visible (UV-Vis) spectroscopic and chromatographic characterization of **3** was performed by gel permeation HPLC using a Beckman Coulter System Gold Bioessential 125/168 diode array detection system (Beckman Coulter, Fullerton, CA) equipped with an in-line Jasco FP-2020 fluorescence detector (Tokyo, Japan). Briefly, 0.020 mg of **3** dissolved in*N,N*-dimethylformamide (DMF, Sigma Biotech grade 99.9+%) was chromatographed with a stationary phase of PLgel MIXED-A column (300 mm×7.5 mm) (Polymer Laboratories, Amherst, MA) and a DMF mobile phase at 1 mL/min at ambient temperature. The extent of DOTA substitution per mass of CNT in compound **3** was determined using the spectrophotometric method [13] of Dadachova *et al.* The number of primary amines that remained after appending the DOTA-NCS was again assayed using the Sarin assay [12]. The mass of CNT was determined using UV-Vis spectroscopy (at 600 nm) to measure the CNT concentration from the linear region of a standard curve of absorbance at 600 nm versus different concentrations of CNT.

Yttrium-86 was produced in the Memorial Sloan-Kettering Cancer Center cyclotron core facility*via* the (p,n) nuclear reaction on an enriched strontium-86 target, and chemically separated from the target using ion chromatographic techniques [14], [15]. Activity was assayed in a Squibb CRC-15R Radioisotope Calibrator (E.R. Squibb and Sons, Inc., Princeton, NJ) set at 711 and dividing the displayed activity value by 2.

The ^{86}Y-CNT construct (Figure 1, compound **4**) was prepared by adding 296 MBq (8 mCi) of acidic ^{86}Y chloride to 0.150 mg of CNT-DOTA (compound **3**) (10 g/L) in MFW and 0.050 mL of 3 M ammonium acetate (Aldrich Chemical Co., Milwaukee, WI) to yield a pH 5.5 solution. The reaction mixture was heated to 60°C for 30 min. and then purified by size exclusion chromatography using a P6 resin (BioRad) as the stationary phase and 1% human serum albumin (HSA, Swiss Red Cross, Bern, Switzerland) in 0.9% NaCl (Abbott Laboratories, North Chicago, IL) as the mobile phase.

The control construct was a mixture of 86Y-DOTA and CNT-amine (compound **2**). This mixture was prepared by the adding 37 MBq (1 mCi) of acidic 86Y chloride to 0.5 mg (10 g/L in MFW) of 1,4,7,10-tetraazacyclododecane-1,4,7,10-tetraaceticacid (DOTA, Macrocyclics, Inc.) and 0.050 mL of 3M ammonium acetate to yield a pH 5.5 solution. The reaction mixture was heated to 60°C for 30 min. and then purified by size exclusion chromatography as described above. 0.150 mg of **2** in MFW was added to the 86Y-DOTA product and mixed.

A small aliquot of each final product was used to determine the radiochemical purity by ITLC-SG using silica gel impregnated paper (Gelman Science Inc., Ann Arbor, MI). The paper strips were developed using two different mobile phases. Mobile phase I was 10 mM EDTA and II was 9% NaCl/10 mM NaOH. The strips were spotted, developed, dried and counted intact using an Ambis 4000 gas ionization detection system (Ambis Inc., San Diego, CA).

Synthesis and Characterization of ^{111}In-CNT

The analogous ^{111}In-CNT construct (compound **5**) was prepared by adding 111 MBq (3 mCi) of acidic ^{111}In chloride (Perkin Elmer, N. Bellerica, MA) to 0.150 mg of **3** (10 g/L) in MFW and 0.050 mL of 3 M ammonium acetate to yield a pH 5.5 solution. The reaction mixture was heated to 60°C for 30 min. and then purified by size exclusion chromatography using a P6 resin as the stationary phase and 1% HSA in 0.9% NaCl as the mobile phase. A small aliquot of the final product was used to determine the radiochemical purity by ITLC-SG using the methods described above. Combined spectroscopic, radiographic, and chromatographic characterization of the CNT construct was performed by reverse phase HPLC, using a Beckman Coulter System Bioessential 125/168 diode array detection instrument equipped with an in-line γ-RAM Model 3 radioactivity detector (IN/US, Tampa, FL). There was a delay of 0.3 min. that corresponded to the time to transit from the diode array detector to the downstream radionuclide detector. Compound **5** (0.010 mg) was analyzed using a Gemini (Phenomenex, Torrence, CA) 5u reverse phase C18 column (250×4.6 mm) to chromatograph these molecules with a 0 to 100% gradient of 0.1M tetraethylammonium acetate (Aldrich), pH 6.5 and acetonitrile (Aldrich) at a flow rate of 1 mL/min.

Experimental Design

Ten male athymic nude mice (Taconic, Germantown, NY), 10–12 weeks old, were separated into three groups. Group I (n = 4) received an intravenous (i.v.) injection of 6.7 MBq (0.18 mCi) and 0.012 mg **4** in 0.20 mL *via* the retroorbital sinus. Group II (n = 3) received an intraperitoneal (i.p.) injection of 6.7 MBq and 0.012 mg **4** in 0.20 mL. Group III (n = 3) received an intravenous (i.v.) injection of 13.3 MBq (0.36 mCi) 86Y-DOTA+0.015 mg of **2** in 0.20 mL *via* the retroorbital sinus. All animals were imaged by PET on day 1 at 3 hours post-injection; group I and II animals were imaged by CT at this time as well, while still under anesthesia. On day 2, 24 hours post-injection, Group I and II animals were imaged by PET. For all *in vivo* experiments housing and care were in accordance with the Animal Welfare Act and the Guide for the Care and Use of Laboratory Animals. The animal protocols were approved by the Institutional Animal Care and Use Committee (IACUC) at Memorial Sloan-Kettering Cancer Center.

CT Imaging

CT imaging was performed using the CT component of the X-SPECT (Gamma Medica, Northridge, CA) a dedicated small-animal SPECT-CT scanner for non-invasive, ultra-high-resolution imaging *in vivo* of single-photon-emitting radiotracers and ultra-high-resolution CT scans for anatomic registration. The imaging times for the CT studies were 10 min. with a resolution of ≈0.100 mm. These CT imaging studies require animals to be fully anesthetized using isofluorane anesthesia.

Micro-PET Imaging

The microPET Focus™ 120 (CTI Molecular Imaging, Inc., Knoxville,

TN) is a dedicated small-animal scanner for imaging PET radiotracers. An energy window of 350–750 keV, a coincidence timing window of 6 nsec, and an acquisition time of 10–20 min. were used. The resulting list-mode data were sorted into 2D histograms by Fourier re-binning and transverse images reconstructed by filtered back-projection into a 128×128×95 matrix. The reconstructed spatial resolution is 2.6 mm full-width half maximum (FWHM) at the center of the field of view. The image data were corrected for non-uniformity of response of the microPET™ (i.e. were normalized), deadtime count losses, physical decay to the time of injection, and the ^{86}Y positron branching ratio but no attenuation, scatter, or partial-volume averaging correction was applied. An empirically determined system calibration factor (i.e. µCi/mL/cps/voxel) for mice was derived by imaging a mouse-size cylinder containing ^{18}F uniformly dispersed in water and used to convert voxel count rates to activity concentrations. The resulting image data were then normalized to the administered activity to determine by region-of-interest analysis the percent of the injected dose per gram (%ID/g) of tissue corrected for radioactive decay to the time of injection.

Data Analysis

AsiPRO VM 5.0 software (Concorde Microsystems, Knoxville, TN) was used to perform image and region of interest (ROI) analyses with the PET and CT datasets. For the ROI analyses, a minimum of 3 regions per tissue per animal were collected and the average %ID/g and standard deviation determined. The average %ID/g tissue per animal was used to determine an average %ID/g per tissue per group and the standard deviation within the group values was calculated. Standards of each injected formulation were counted to quantify the %ID/g. Unpaired two-tailed t-tests were performed to assess temporal differences of tissue activity. Prism software (Graphpad Software, Inc., San Diego, CA) was used for statistical analyses and plotting data. TEM image analysis was performed using ImageJ software (NIH, http://rsb.info.nih.gov/ij/). AFM image analysis was carried-out using Nanoscope II software from Digital Instruments.

Biodistribution Data

Animals were sacrificed at 24 h post-injection and the kidneys, liver and spleens were harvested according to approved IACUC institutional protocols. Tissue samples were weighed and counted in a Packard Cobra γ-counter (Packard Instrument Co., Inc., Meriden, CT) using a 315–435 KeV window. Standards of each injected formulation were counted to determine the %ID/g.

Blood Clearance and Excretion of ^{111}In-CNT

In another biodistribution study, 0.007 mg of **5** (specific activity 37 GBq/g (1Ci/g)) were injected i.v., *via* the retroorbital sinus, per BALB/c mouse (female, 8–12 weeks old, Taconic Farms, NY) and the mice were divided into 5 groups of n = 3 per group. One group of mice was sacrificed at each of the following time points 1, 24, 96, 216 and 360 hours and tissue samples including blood, brain, lung, heart, adipose tissue, liver, kidney, spleen and femur were harvested, weighed and counted using a Packard Cobra Gamma counter using the 15–550 KeV window. The %ID/g was determined by measuring the activity of an aliquot of each construct injectate. Urine samples were collected from mice 1 hour post injection. Activity was detected and the urine samples were analyzed using the ITLC method described above.

RESULTS

The synthesis, radiolabeling, and characterization of the CNT constructs Pristine CNT were purified and shortened by digestion in dilute nitric acid and sonication. The product of this process, **1**, was carboxy-functionalized CNT. The Raman spectrum of this product is shown in Figure 2a. The radial breathing mode (RBM) feature at 200 cm−1 yields a CNT diameter of 1.1 nm and further confirmation of the product identity is provided by the 1340 cm−1 disorder band (D band) and the sharp 1556 and 1584 cm−1 tangential mode (G band) resonances [16], [17]. These CNT (compound **1**) were functionalized and solubilized by the covalent sidewall attachment of pendant primary amines to yield compound **2**. The Raman spectrum of **2** (Figure 2b) was characterized by the loss of the RBM feature and a rising baseline due to increasing luminescence resulting from the enhanced dispersion of this product in solution [18]. The amine content of the construct was quantified using the Sarin assay [12], which yielded a value of 1.76 mmol amine/g CNT. Subsequently, these amines were reacted with DOTA-NCS to yield the CNT-DOTA construct (compound **3**). Raman characterization of **3**(Figure 2c) confirmed the enhanced solubility (~10 g/L) of this construct by the rising baseline[18], which partially masks the D and G bands. Transmission electron and atomic force microscopic images (Figure 3) of **3** are presented to demonstrate the composition, identity and purity of the DOTA-functionalized CNT product. Figure 3a displays the results of an image analysis of the TEM data of a sampling of 22 'unbundled' **3** in Figure 3b. This data shows that the mean±standard deviation CNT length distribution of **3** was 42±17 nm (n = 22). The AFM and TEM images of **3** in Figures 3c and d, respectively, show bundles of **3** with varying thickness (approximately 1–20 nm) and lengths and the absence of carbonaceous or nanoparticulate contaminants. A gel permeation HPLC chromatograph of **3** is shown in Figure 4 with a major absorbance peak at 11.7 min and the corresponding UV-Vis absorbance spectrum of that peak demonstrating the characteristic CNT spectral feature. This gel permeation chromatograph further demonstrates the purity of **3**, notably by the absence of the later eluting amorphous carbon and nanoparticulate species [19], which are highly fluorescent species [18] and that were not detected using the in-line fluorescence detector. The DOTA content of **3** was determined using a lead arsenazo assay [13], which yielded a value of 0.30 mmol DOTA/g CNT; in addition **3** still contained some unreacted amines assayed to be 0.27 mmol amine/g CNT. Construct **3** was labeled with "no carrier added" 86Y chloride [14], [15] and purified by size exclusion chromatography. The radiochemical purity of compound **4** was determined to be 90% by ITLC methods and had a specific activity of 555 GBq/g (15 Ci/g). The control construct was a mixture of 86Y-DOTA+**2**. The 86Y-DOTA was 95% radiochemicaly pure as determined by ITLC-SG.

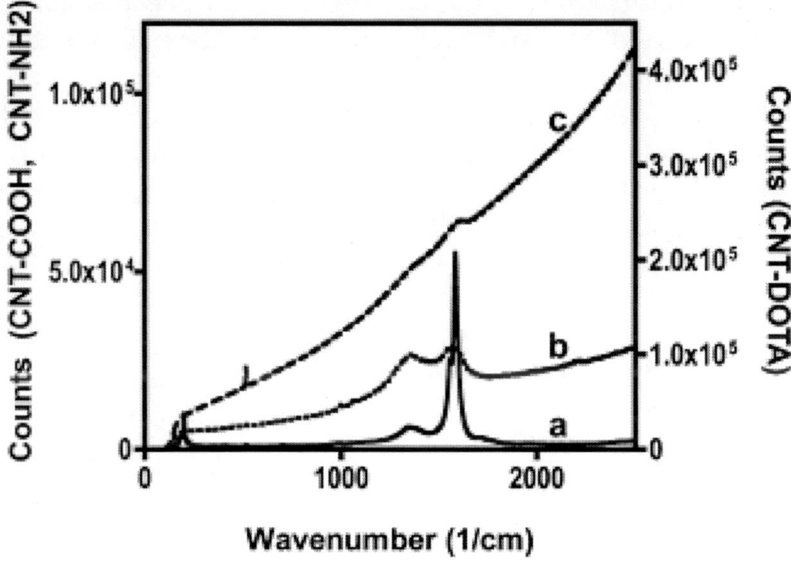

Figure 2. Raman spectra measured with a 488 nm laser.

(a) acid-oxidized and purified CNT, compound 1; (b) sidewall amine-functionalized CNT, compound 2; (c) DOTA-functionalized CNT, compound 3. The spectra for 1 and 2 are plotted on the left axis (0-120,000), and 3 on the right axis (0–450,000).

An ^{111}In-CNT construct (compound 5) was similarly prepared and purified. This radiochemical analog was used in ancillary animal experiments to measure blood clearance and excretion *in vivo* over an extended period of time due to it›s longer half-life. ^{111}In has been extensively utilized as a surrogate radionuclide for yttrium chemistry in pre-clinical and clinical studies. Compound 5 also served to validate the ITLC-SG methods for determining radiochemical purity. The ^{111}In-CNT product was determined to be 90% by the ITLC methods described above, similar to the observed ITLC-SG purity of 4. A reverse phase HPLC analysis of 5 confirmed the co-elution of CNT and radioactivity (Figure 5). The diode array detected a sharp absorbance peak at 12.7 minutes (Figure 5b) that was attributed to the ^{111}In-CNT product and confirmed by the characteristic CNT spectral signature (Figure 5c). The corresponding radioactivity trace (Figure 5a) revealed a sharp peak at 13.0 minutes that contained 90% of the eluted radioactivity activity and after correction for the delay between detectors, was assigned to the ^{111}In-CNT product. Two uniquely different chromatographic methods were developed and yielded the same result for the radiochemical purity - the reverse phase HPLC method and the isocratic ITLC method. The reverse phase HPLC data also correlated the radioactivity with the CNT, thus validating the use of ITLC to rapidly and quantitatively determine radiochemical purity and identity of these constructs.

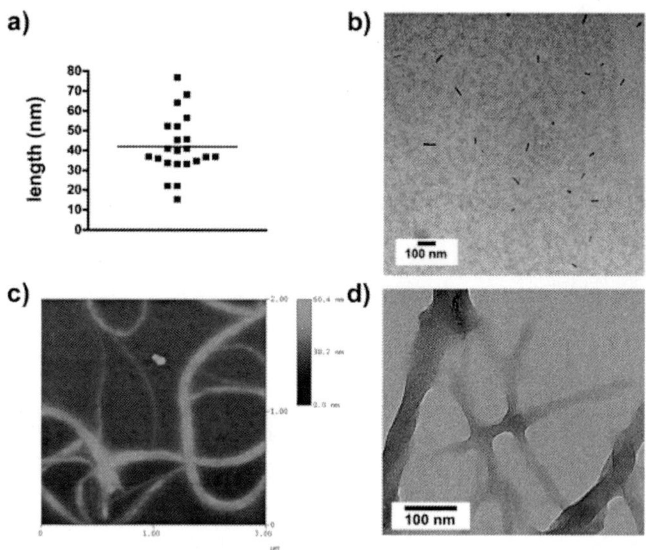

Figure 3 Microscopic images of 3 provide data to evaluate the (a) size distribution of 22 individual DOTA-CNT obtained from (b) a representative TEM image of solubilized 3 (scale bar = 100 nm); (c) AFM image of the height (scale bar = 0–60 nm) and (d) TEM image (scale bar = 100 nm) of representative DOTA-CNT bundles.

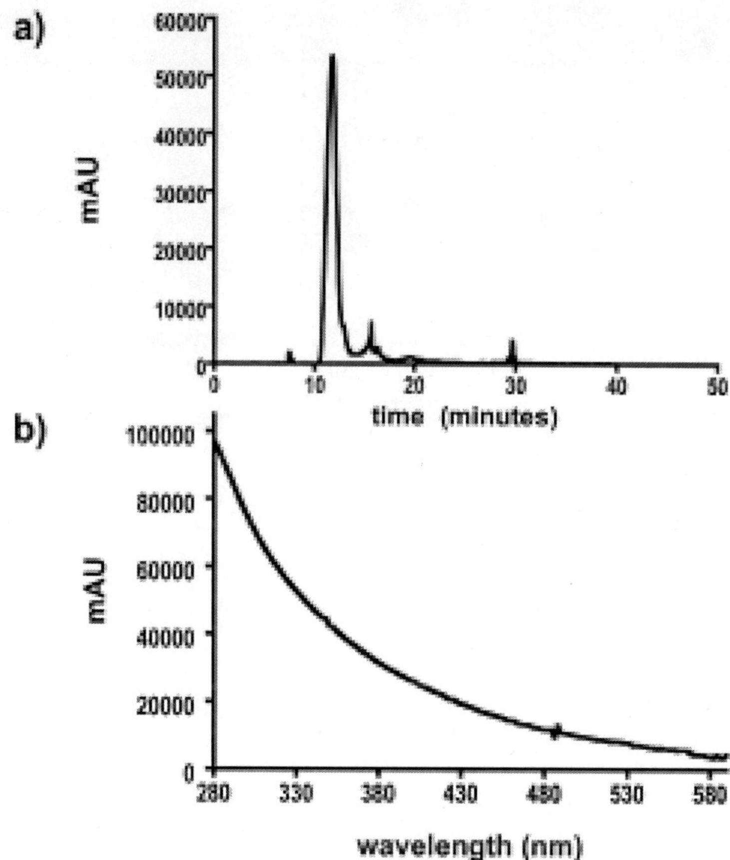

Figure 4. Gel permeation HPLC chromatograph of compound 3 (0.020 mg) dissolved in DMF on a stationary phase of PLgel MIXED-A column with DMF mobile phase at 1 mL/min at ambient temperature.

(a) Chromatograph is shown with absorbance at 330 nm. (b) The UV-VIS absorbance spectrum of the peak at 11.7 min.

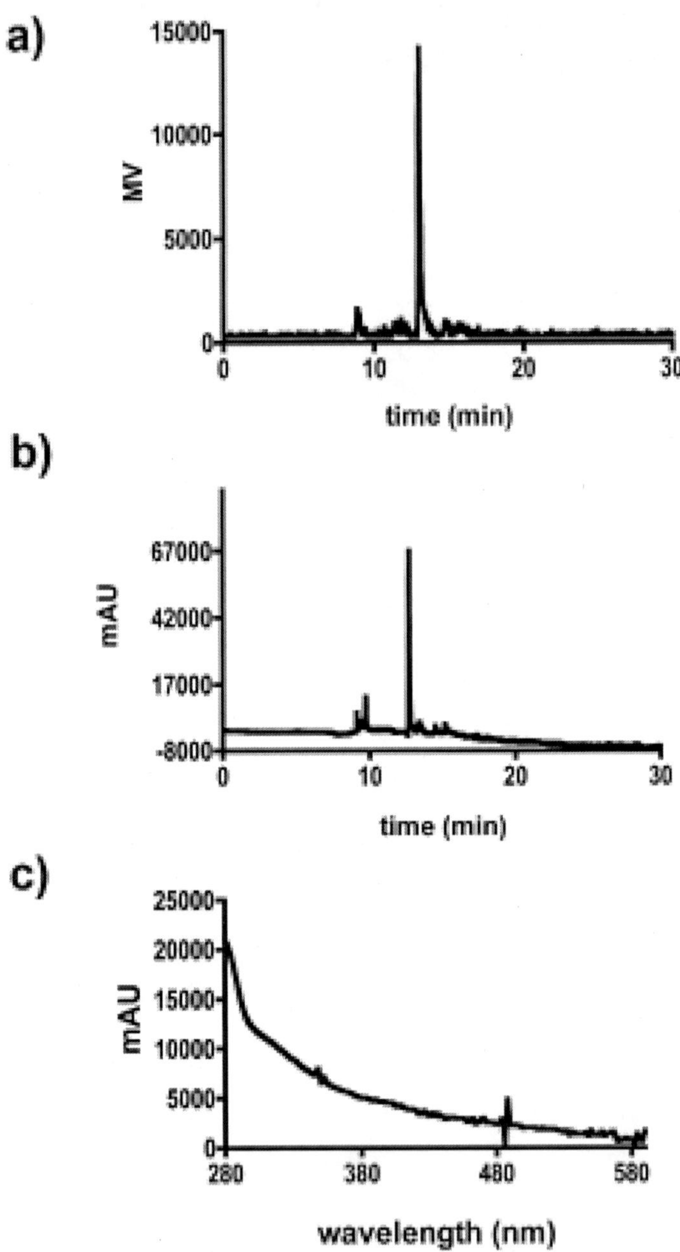

Figure 5. Reverse phase HPLC chromatograph of ^{111}In-CNT, compound 5 (0.010 mg); (a) Radionuclide trace of 5; (b) Corresponding chromatograph with absorbance at 350 nm (***n.b.***, there was a 0.3 min delay between the detectors); (c) The UV-VIS absorbance spectrum of the peak at 12.7 minutes in panel b.

The biodistribution of the radiolabeled construct and the control construct

The ten mice were separated into three groups, each mouse in groups I and II received 0.012 mg of **4** labeled with 6.7 MBq of ^{86}Y, administered i.v. or i.p., respectively. Animals in group III (control group) each received a physical mixture of 13.3 MBq of ^{86}Y-DOTA and 0.015 mg of **2***via* i.v. injection. We evaluated the biodistribution of **4** *in vivo* by serial PET and then performed anatomic imaging using a dedicated small-animal CT scanner. Software fusion of the resulting PET and CT images was carried out to combine the images. The imaging results were confirmed by sacrificing the animals and harvesting, weighing, and counting the activity in the organs.

PET images of animals in groups I and II at 3 hours post-injection clearly demonstrated accumulation of activity in the kidneys, spleen and liver (Figure 6). The kidney images showed uptake primarily in the renal cortex, while the renal medulla was devoid of activity. There was no blood-pool activity observed in these images. In group I, the ^{86}Y-CNT construct appeared to avidly localize in the spleen and liver. In group II, spleen and liver uptake was significantly lower and a diffuse activity was evident from the i.p. cavity suggesting a relatively slow egress from the i.p. compartment into the vascular compartment. After 24 hours there was no significant difference in the accumulated activity in the spleen and liver within each group compared to the 3 hour time-point. Additionally, the kidneys of the animals in groups I and II had begun to clear the constructs. These data indicate that the clearance from the spleen and liver is slower than from the kidneys. The radioactivity in the liver of the group I animals was still significantly higher than the group II animals at both time-points evaluated. As expected for a stably chelated metal-ion species with a relatively small molecular weight, the control mixture cleared completely from group III within 3 hours. The imaged radioactivity was located primarily in the bladder (~1 %ID/g at 3 hours) and was consistent with urinary excretion of small molecules (data not included).

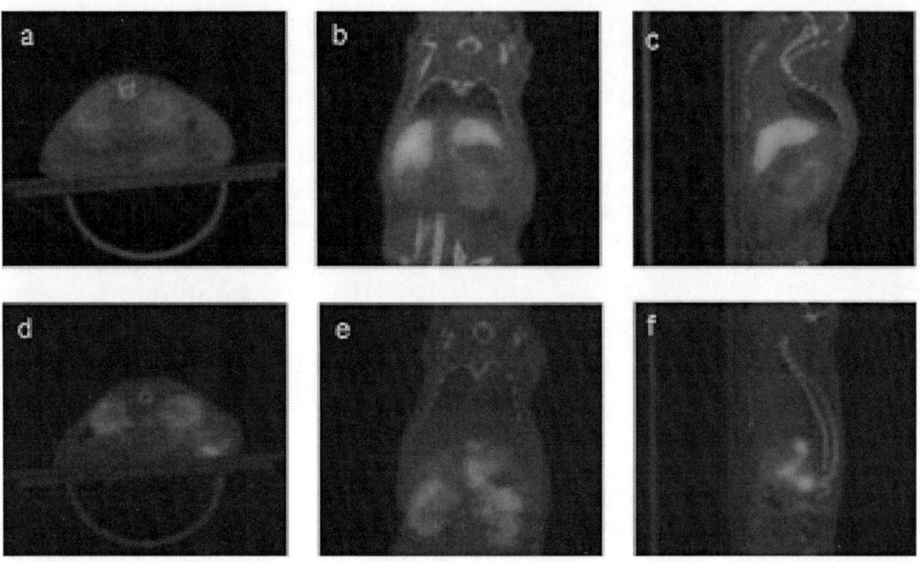

Figure 6. Fused PET and CT images at 3 hours post-injection of ^{86}Y-CNT, compound 4, showing a transverse (z-direction, images a and d), a coronal (y-direction, images b and e), and a sagittal (x-direction, images c and f) slice of one representative animal from group I (i.v., a, b and c) and one from group II (i.p., d, e and f).

We performed region of interest (ROI) analysis of the PET images in each of the group I and II animals to extract a decay-corrected activity concentration in various organs. Data (Figure 7a) are reported as the mean±standard deviation %ID/g. The kidneys had 8.30±0.92 and 8.50±1.05; the liver had 17.8±3.95 and 7.54±1.38; the spleen had 14.3±2.0 and 9.7±1.3; and the bone (femur and spine) had 2.26±0.14 and 1.44±0.23 %ID/g at 3 hours post-injection in the group I and II animals, respectively.

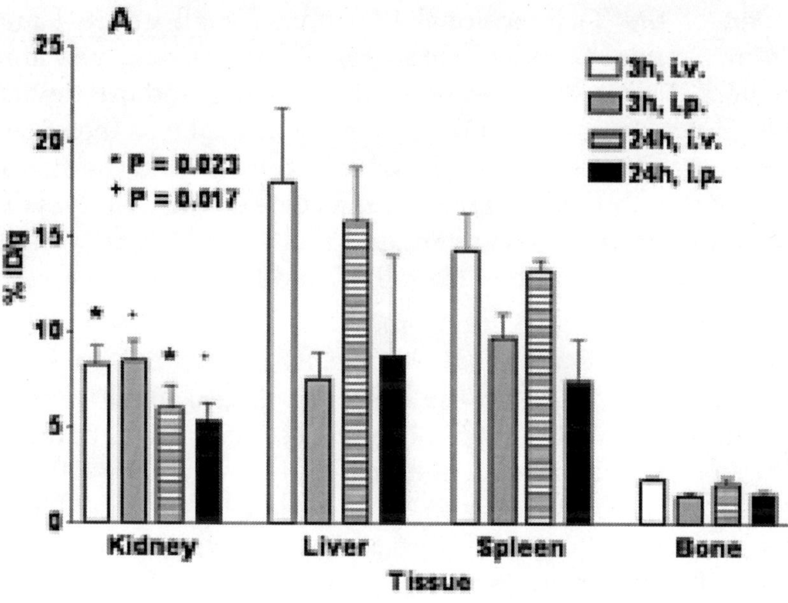

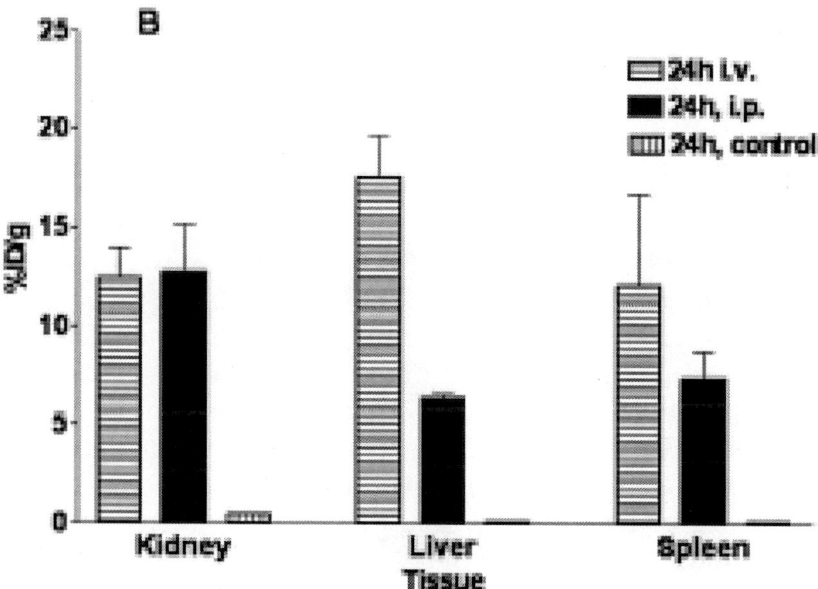

Figure 7. Biodistribution data obtained from the PET imaging experiment and from tissue harvest.

(a) Plot of the 2-dimensional ROI data for all group I and II animals at 3 and 24 hours. Three regions per tissue per animal were collected and the average %ID/g and standard deviation determined. The average %ID/g tissue per animal was then used to determine an average %ID/g per tissue per group and the standard deviation within the group values was also calculated. (b) Plot of the 24 hour biodistribution data obtained from tissue harvest, weighing and counting of organs from group I, II, and III mice.

At 24 hours post-injection, the kidneys had 5.96±1.23 and 5.42±0.85; the liver had 15.8±2.90 and 8.83±5.17; the spleen had 13.2±0.60 and 7.51±2.12; and the bone had 2.02±0.36 and 1.56±0.17 %ID/g in the group I and II animals, respectively. The accumulated activity in the kidneys cleared significantly in group I ($P = 0.023$) and group II ($P = 0.017$) over this 21-hour period. However, statistical analysis showed that the activity accumulated in the spleen, liver and bone was not clearing over this period of time.

Residual ^{86}Y-CNT activity that was slowly exiting the i.p. compartment yielded poorer image contrast in the group II mice. Interestingly, the construct activity that did exit the i.p. cavity had the same kidney uptake as the i.v. administered construct at both timepoints, but very different liver and spleen accumulation. The %ID/g activity in the livers of group II animals is 42% of that in the group I animals at 3 hours post-injection, and 56% of that in the group I animals at 24 hours post-injection. It appeared that the equilibration process is slow and the liver activity increases with time in the group II animals.

Biodistribution data (Figure 7b) obtained from tissue harvest at 24 hours was in good agreement with the ROI data. The spleen showed the same differential accumulation as the liver as a function of i.p. *versus* i.v. administration. Group III control animals showed no organ-specific uptake of ^{86}Y with only small amounts of activity (0.1 to 1 %ID/g) in the kidney, liver and spleen. The post-mortem biodistribution data and the 3-hour PET images showed this control mixture rapidly cleared from the group III mice.

The whole-body PET images indicated that the major sites of accumulation of activity resulting from the administration of ^{86}Y-CNT were the kidney, liver, spleen, and to a much less extent the bone. The % ID/organ values for kidney, liver, and spleen, 24 hours post-i.v. injection, were 5.96±1.20, 15.2±1.54, and 0.82±0.04, respectively. The corresponding % ID/organ values for kidney, liver, and spleen, 24 hours post- i.p. injection, were 5.48±0.61, 4.46±1.08, and 0.60±0.09, respectively. Whole-body activity was decreasing over this 24 hour period of time, as we could account for only about 20% of the total i.v. administered activity in the group I animals and about 11% of the total activity in the group II animals. In another biodistribution experiment, ^{111}In-CNT was injected into mice and ITLC analysis of urine samples obtained 1 hour after injection was consistent with ^{111}In-CNT being excreted.

The clearance of radiometal-labeled CNT constructs from the blood compartment is rapid. This was observed in this study (no blood pool activity at 3 h) and quantitatively measured in another biodistribution study using the longer half-lived ^{111}In. Briefly, compound 5 was injected into mice using the retroorbital sinus route and venous blood samples were obtained 1, 20, 96, 216, and 360 hours after injection from groups of three animals per timepoint. The data are shown inFigure 8. After 1 hour 2.76±0.14 %ID/g was still in the blood and after 20 h only 0.41±0.05 %ID/g was in circulation. The experiment was continued for 15 d at which time no activity was found in circulation. Additional ^{111}In-CNT biodistribution data are presented in a study of lymphoma tumor targeting with an antibody-functionalized, radiolabeled carbon nanotube [20]. Tissue harvest indicated that the major sites of ^{111}In accumulation were the kidney, liver, spleen, and to a much less extent the bone, similarly to the ^{86}Y accumulation. ^{111}In cleared the kidneys more rapidly than the spleen and liver.

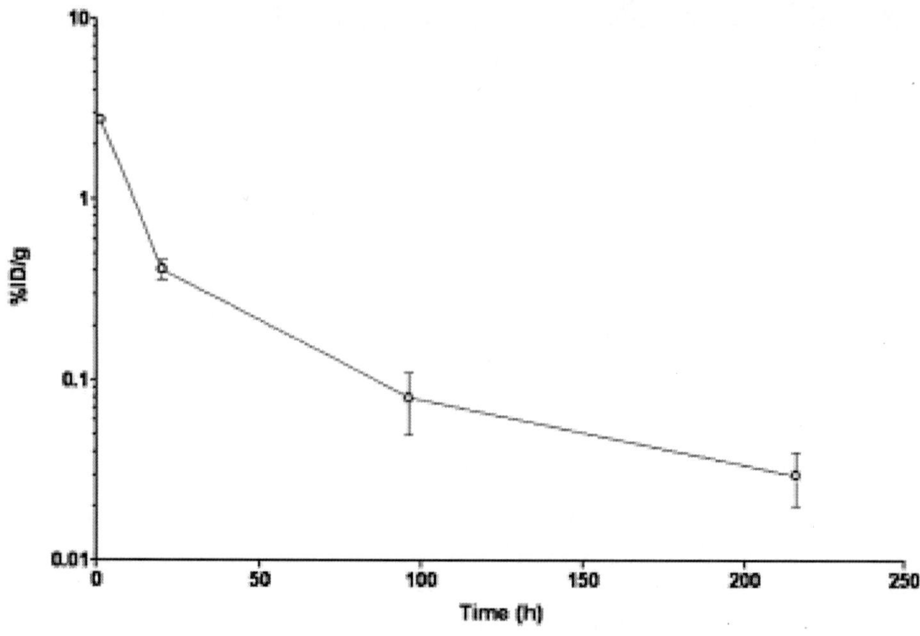

Figure 8. Plot of the blood clearance of ^{111}In-CNT, compound 5, following i.v. injection (***n*** = 3 per group).

DISCUSSION

The interface of CNT with biological systems is a key issue that can be addressed in part by determining the pharmacokinetic profile of these unique materials. Where do functionalized-CNT distribute *in vivo* following administration and how do they clear are the questions that were posed in this study. Published reports of the biodistribution of soluble, sidewall-functionalized, radiolabeled CNT in murine models [6], [7], [20] were very different. The first two biodistributions were based on tissue harvest and did not contain an imaging component. Wang et al. [6] used ^{125}I-CNT (*l*~300 nm) where the CNT were hydroxyl-functionalized. They reported the ^{125}I cleared the blood and accumulated in the stomach, kidneys and bone; stomach accumulation is not unusual for free ^{125}I. Bone and kidney had accumulated activity that persisted for up to 6 days. 80% of the activity cleared the animals over an 11 day period. They did not report notably different tissue accumulations when the construct was administered i.v. or i.p. Singh et al. [7] used CNT (*l*~300–1000 nm) that were amine-functionalized, modified with DTPA anhydride, and then labeled with ^{111}In. The ^{111}In cleared the blood compartment rapidly (within 3 hours) and surprisingly there was no significant tissue uptake. McDevitt et al. [20]reported the selective targeting of tumor *in vivo* using tumor-specific antibodies that were covalently-appended to a soluble, ^{111}In-radiolabeled CNT construct relative to controls of a i) non-targeting antibody-appended CNT construct; ii) a non-targeting CNT construct, and iii) antibody alone. Clearance of the non-targeting, radiolabeled CNT construct from the blood compartment was rapid. Tissue harvest indicated that the major sites of ^{111}In accumulation were the kidney, liver, spleen, and to a much less extent the bone. ^{111}In cleared the kidneys more rapidly than the spleen and liver. Radioactive CNT was in the urine samples collected from mice 1 hour post injection. The covalent attachment of antibodies to the CNT scaffold dramatically altered the kidney biodistribution and pharmacokinetics relative to the non-targeting, radiolabeled CNT construct.

Small globular molecules (<25 kDa) are typically filtered by the kidney but may also be reabsorbed subsequently in the tubule. Large globular molecules usually bypass clearance through the kidney, CNT may not behave similarly due to their large aspect ratio, unlike globular proteins. Singh *et al.* [7] observed rapid and complete renal clearance of their injected material, which is difficult to explain given the 25 kDa cut-off of the glomerulus and the 0.1 to 1 mDa (approximate) weight range of the CNT materials.

Compound **3**, the DOTA-CNT starting material, used to prepare the radiolabeled constructs **4**and **5**, was microscopically, spectroscopically, and chromatographically characterized to confirm the identity and purity of this reagent. These were "short" tubes with mean lengths of 42±17 nm (n = 22) resulting from the acid digestion and sonication processing steps [21]–[23]. The AFM and TEM images of **3** show bundles of varying thickness and lengths with no carbonaceous or particulate contaminants. Gel permeation chromatography of **3** identified a major absorbance peak that had the spectral signature of CNT and which did not exhibit any carbonaceous or particulate contaminants. Compounds **4** and **5** were characterized by traditional radiochemical ITLC methods which were validated using radio-HPLC methods. The radioactivity (^{86}Y or ^{111}In) was clearly associated with the CNT, which is not surprising given the stability of the DOTA-metal-ion complex and the thiourea linkage [24] between the DOTA and the CNT. Chelates such as DTPA anhydride were significantly less stable than DOTA or backbone-modified DTPA moieties [25]–[27].

Compounds **4** and **5**, that we report herein, were fully water-soluble and were stable in the 1% human serum albumin (HSA) injection vehicle and stable in biological fluids [20]. We were not surprised that high molecular-weight, water-soluble compounds that are CNT-based were accumulated in the liver and spleen and also in the kidney. Previous work [6], [20] also observed accumulation of their CNT constructs in the kidneys. Hydrophilic fullerenes were reported [28], [29] to partition into liver and spleen, but exhibit little-to-no accumulation in kidney, presumably because of their smaller size and shape. It was surprising that the constructs reported by Singh *et al.* clear the mouse entirely within a few hours and do not accumulate anywhere.

A pharmacokinetic report [8] of unmodified CNT (l~300 nm) suspended in a surfactant and injected i.v. into rabbits employed the inherent near-infrared fluorescence of CNT for detection in blood and tissue samples. They report a rapid blood clearance (half-life of 1.0 h) and were only able to detect CNT in the liver, but not in any other tissue sample examined (0.020 mg/kg was administered). The materials that they examined were unmodified, insoluble CNT and these may exhibit very different properties than the modified, soluble CNT constructs used in this and the other three studies [6], [7], [20]. Others have coated CNT with non-covalently attached radiolabeled- and RGD-labeled surfactants and administered (0.05 mg/kg) to mice [30]. This type of construct is more similar to the unmodified CNT [8] in that the CNT was dispersed in a detergent solute to facilitate solubilization. These molecules were fundamentally very different than the robust, covalently assembled constructs reported herein and elsewhere [6], [7], [20], since the radiolabel and the RGD were not attached to the CNT. Unfortunately, they did not report [30] the biodistribution of the radiolabeled surfactant alone, so that one can compare its biodistribution to that of the construct nor of a control construct (e.g., labeled with a non-specific control peptide sequence). Thus the biodistribution can not be inferred with certainty.

In this biodistribution study we performed an *in vivo* imaging study of soluble, sidewall-functionalized, radiolabeled CNT constructs in mice utilizing PET. By employing PET imaging (and tissue harvest), we were able to measure accumulation and clearance in groups of living animals over time by observing each subject. The radioactivity from these compounds also accumulated in bone and kidney, was cleared rapidly from the blood, and was excreted in the urine. The spleen and liver accumulation, and the differential clearance from the i.p. cavity, may result from a different hydrophobicity of the hydroxylated [6] *versus* chelate/amine functionalized CNT. While our CNT constructs were processed similarly to those in reference 7, in that paper the isotope completely cleared the blood and body rapidly.

We speculated that the CNT most likely cleared *via* i) longitudinal filtration [31] through the renal glomeruli, which has a pore size of approximately 30 nm diameter or ii) active secretion into the tubular lumen. Accumulation of CNT in the kidneys, in particular in the renal cortex region where filtration occurs, was clearly shown in Figure 6a-c. The effect of molecular shape, charge and size on kidney clearance has been investigated using different macromolecular solutes [32], [33]. The frictional ratio of the molecule plays an important role as does the shape in crossing the glomerular barrier. The negatively charged capillary endothelium, anionic glomerular basement membrane, and the anionic podocyte cell coating might provide a charge gradient that facilitates transit of amine-functionalized CNT from the blood to the urine [31]. Differences in CNT construct purity, chemistry, charge, and flexibility could alter the biodistribution. The covalent attachment of antibody molecules to the CNT dramatically altered the kidney biodistribution and pharmacokinetics compared to ^{111}In-CNT in tumor-bearing*versus* non-tumor-bearing mice [20]. PET/CT provides a comprehensive measure of *in vivo*behavior to which the physicochemical properties of each construct can be compared.

A comparison of the PET images obtained at 3 and 24 hours demonstrated that the diffuse activity imaged in the i.p. cavity of the group II animal persists over the 24 hour period of the experiment and the liver activity was less than that in the group I animals. This observation was interesting since large globular proteins, for example IgG (~150 kDa), injected into the i.p. cavity of a mouse redistributed rapidly and reached equilibrium in the blood within 5 to 6 hours (data not shown). The mean ^{86}Y-CNT construct size was approximately 100 kDa, but obviously it did not behave as a globular protein. The persistence of radioactivity in the i.p. cavity seemed to minimize liver and spleen accumulation.

The mass amounts of CNT construct that were used per animal in this study (0.6 mg/kg) were in the same range as those that were employed in a study with radio-labeled, targeting-antibody CNT constructs in mice [20]. The amount of activity accumulated in the organs of these animals after 24 hours (The % ID/organ values for kidney, liver, and spleen, post- i.v. injection, were 5.96±1.20, 15.2±1.54, and 0.82±0.04, respectively; the corresponding values post-i.p. injection, were 5.48±0.61, 4.46±1.08, and 0.60±0.09, respectively.) does not pose a problem in translating these constructs to other studies. Further, we have demonstrated that appropriately designed targeting diagnostic agents based on a CNT platform might further alter liver, kidney and spleen accumulation/clearance [20]. The rapid blood clearance of these nanomaterials could prove beneficial when using short half-lived radionuclides in diagnostic applications, as prolonged persistence of blood-pool activity following antibody-construct administration obscures target imaging due to a low signal-to-background ratio. The mechanism of excretion from the kidneys must also be further examined to better understand the construct parameters that affect clearance.

CONCLUSION

The whole-body PET images indicated that the major sites of accumulation of radioactivity resulting from the administration of soluble, sidewall-functionalized ^{86}Y-CNT were the kidney, liver, spleen, and to a much less extent the bone. PET data provides extremely sensitive, quantitative, and functional information that is different from that obtainable with other largely anatomical imaging modalities and CT was performed to confirm anatomical assignments. The presence of activity in the renal cortex where glomerular filtration occurs was clearly shown inFigure 6 and correlates with the radioactivity found in the urine. Furthermore, the rapid blood clearance that was observed can be beneficial in the use of short-lived radionuclides in diagnostic applications by reducing the background activity compared to target tissue activity. The mass amounts of soluble CNT construct that were used per animal (0.6 mg/kg) were 30-fold higher than those applied in a study with un-functionalized CNT [8] and point-out the utility of functionalized-CNT which not only improves solubility but acts as a scaffold that can bear many copies of an appended moiety. The CNT compounds in this study were very short-length and chemically and radiochemically pure. The extensive physical and chemical characterization studies yielded a clear description of the identity and purity of these CNT compounds and proved useful in better understanding their pharmacokinetic profile. Furthermore, the careful radiochemical characterization and experimental controls better correlated the observed pharmacokinetic results with the composition and identity of the material injected.

We believe that the results described herein encourage further study of CNT constructs in vivo. This soluble, sidewall-functionalized CNT platform rapidly cleared the blood compartment and exhibited some distribution to the expected clearance sites. We have also demonstrated the utilization of an antibody-modified-CNT construct in targeting human tumor in a murine model[20]. The emerging field of nano-medicine will find applications for many such CNT-based molecular constructs based upon this synthetic and imaging approach.

ACKNOWLEDGMENTS

The technical assistance of Ms. Valerie Longo and Ms. Cindy Usher.

AUTHOR CONTRIBUTIONS

Conceived and designed the experiments: MM DC DS. Performed the experiments: JJ MM DC RF PZ CV DR JM. Analyzed the data: MM DC DS CV DR JM CB. Contributed reagents/materials/analysis tools: MM DC RF PZ JN DR JM CB. Wrote the paper: MM DS.

REFERENCES

1. Dresselhaus MS, Avouris P (2001) Introduction to carbon materials research. Topics in Applied Physics 80: 1–9.
2. In het Panhuis M (2003) Vaccine delivery by carbon nanotubes. Chem Biol 10: 897–898.
3. Cherukuri P, Bachilo SM, Litovsky SH, Weisman RB (2004) Near-infrared fluorescence microscopy of single-walled carbon nanotubes in phagocytic cells. J Am Chem Soc 26: 15638–15639.
4. Shi Kam NW, Jessop TC, Wender PA, Dai HJ (2004) Nanotube Molecular Transporters: Internalization of Carbon Nanotube-Protein Conjugates into Mammalian Cells. J Am Chem Soc 126: 6850–6851.
5. Pantarotto D, Briand JP, Prato M, Bianco A (2004) Translocation of bioactive peptides across cell membranes by carbon nanotubes. Chem Commun 7: 16–17.
6. Wang H, Wang J, Deng X, Sun H, Shi Z, et al. (2004) Biodistribution of carbon single-wall nanotubes in mice. J Nanosci Nanotechnol 4: 1019–1024.
7. Singh R, Pantarotto D, Lacerda L, Pastorin G, Klumpp C, et al. (2006) Tissue biodistribution and blood clearance rates of intravenously administered carbon nanotube radiotracers. Proc Natl Acad Sci U S A 103: 3357–3362.
8. Cherukuri P, Gannon CJ, Leeuw TK, Schmidt HK, Smalley RE, et al. (2006) Mammalian pharmacokinetics of carbon nanotubes using intrinsic near-infrared fluorescence. Proc Natl Acad Sci U S A 103: 18882–18886.
9. Georgakilas V, Tagmatarchis N, Pantarotto D, Bianco A, Briand JP, et al. (2002) Amino acid functionalisation of water soluble carbon nanotubes. Chem Commun 12: 3050–3051.
10. Pantarotto D, Partidos DC, Graff R, Hoebeke J, Briand JP, et al. (2003) Synthesis, structural characterization, and immunological properties of carbon nanotubes functionalized with peptides. J Am Chem Soc 125: 6160–6164.
11. Von Schulthess GK, Steinert HC, Hany TF (2006) Integrated PET/CT: Current Applications and Future Directions. Radiology 238: 405–422.
12. Sarin VK, Kent SBH, Tam JP, Merrifield RB (1981) Quantitative monitoring of solid-phase peptide synthesis by Ninhydrin reaction. Anal BioChem 117: 147–157.
13. Dadachova E, Chappell LL, Brechbiel MW (1999) Spectrophotometric method for determination of bifunctional macrocyclic ligands in macrocyclic ligand-protein conjugates. Nucl Med Bio l26: 977–982.
14. Finn RD, McDevitt M, Ma D, Jurcic J, Scheinberg D, et al. (1999) Low Energy Cyclotron Production and Separation of Yttrium-86 for Evaluation of Monoclonal Antibody Pharmacokinetics and Dosimetry. In: Duggan JL, Morgan IL, editors. Application of Accelerators in Research and Industry. Woodbury, NY: American Institute of Physics. pp. 991-993:

15. Lovqvist A, Humm JL, Sheikh A, Finn RD, Koziorowski J, et al. (2001) PET imaging of (86)Y-labeled anti-Lewis Y monoclonal antibodies in a nude mouse model: comparison between (86)Y and (111)In radiolabels. J Nucl Med 42: 1281–1287.

16. Yu Z, Brus L (2001) Rayleigh and Raman scattering from individual carbon nanotube bundles. J. Phys. Chem. B 105: 1123–1134.

17. Itkis ME, Perea DE, Jung R, Niyogi S, Haddon RC (2005) Comparison of analytical techniques for purity evaluation of single-walled carbon nanotubes. J. Am Chem. Soc. 127: 3439–3448.

18. Lin Y, Zhou B, Martin RB, Henbest KB, Harruff BA, Riggs JE, Guo Z-X, Allard LF, Sun Y-P (2005) Visible luminescence of carbon nanotubes and dependence on functionalization. J. Phys. Chem. B 109: 14779–14782.

19. Zhao B, Hu H, Niyogi S, Itkis ME, Hamon MA, Bhowmik P, Meier MS, Haddon RC (2001) Chromatographic purification and properties of soluble single-walled carbon nanotubes. J. Am. Chem. Soc. 123: 11673–11677.

20. McDevitt MR, Chattopadhyay D, Kappel BJ, Jaggi JS, Schiffman SR, et al. (2007) Tumor targeting with antibody-functionalized, radiolabeled carbon nanotubes. J. Nucl. Med. 48: 1180–1189.

21. Huang W, Lin Y, Taylor S, Gaillard J, Rao AM, Sun Y-P (2002) Sonication-assisted functionalization and solubilization of carbon nanotubes. Nano Letters 2(3): 231–234.

22. Hu H, Zhao B, Itkis ME, Haddon RC (2003) Nitric acid purification of single-walled carbon nanotubes. J. Phys. Chem. B 107: 13838–13842.

23. Arnold K, Hennrich F, Krupke R, Lebedkin S, Kappes MM (2006) Length separation studies of single walled carbon nanotube dispersions. Phys Stat. Sol. (B) 243(13): 3073–3076.

24. McDevitt MR, Ma D, Lai LT, Simon J, Borchardt P, et al. (2001) Tumor therapy with targeted atomic nano-generators. Science 294: 1537–1540.

25. Esteban JM, Schlom J, Gansow OA, Atcher RW, Brechbiel MW, et al. (1987) New method for the chelation of indium-111 to monoclonal antibodies: biodistribution and imaging of athymic mice bearing human colon carcinoma xenografts. J Nucl Med 28: 861–870.

26. Ruegg CL, Anderson-Berg WT, Brechbiel MW, Mirzadeh S, Gansow OA, et al. (1990) Improved in vivo stability and tumor targeting of bismuth-labeled antibody. Cancer Res 50: 4221–4226.

27. Brouwers AH, van Eerd JE, Frielink C, Oosterwijk E, Oyen WJ, et al. (2004) Optimization of radioimmunotherapy of renal cell carcinoma: labeling of monoclonal antibody cG250 with 131I, 90Y, 177Lu, or 186Re. J Nucl Med 45: 327–337.

28. Bullard-Dillard R, Creek KE, Scrivens WA, Tour JM (1996) Tissue sites of uptake of C-14-labeled C-60. Bioorg Chem 24: 376–385.

29. Cagle DW, Kennel SJ, Mirzadeh S, Alford JM, Wilson LJ (1999) In vivo studies of fullerene-based materials using endohedral metallofullerene radiotracers. Proc Natl Acad Sci U S A 96: 5182–5187.

30. Lui Z, Cai WB, He L, Nakayama N, Chen K, et al. (2007) In vivo biodistribution and highly efficient tumour targeting of carbon nanotubes in mice. Nature Nanotech 2(1): 47–52.
31. Lee CC, MacKay JA, Frechet JM, Szoka FC (2005) Designing dendrimers for biological applications. Nat Biotechnol 23: 1517–1526.
32. Ohlson M, Sorensson J, Lindstrom K, Blom AM, Fries E, et al. (2001) Effects of filtration rate on the glomerular barrier and clearance of four differently shaped molecules. Am J Physiol Renal Physiol 281: F103–F113.
33. Moffat DB (1981) New ideas on the anatomy of the kidney. J Clin Pathol 34: 1197–1206.

Chapter 6

INTRACELLULAR NEURAL RECORDING WITH PURE CARBON NANOTUBE PROBES

Inho Yoon, Kosuke Hamaguchi, Ivan V. Borzenets, Gleb Finkelstein, Richard Mooney Mail, Bruce R. Donald Mail

Department of Electrical and Computer Engineering, Duke University, Durham, North Carolina, United States of America

Department of Neurobiology, Duke University Medical Center, Durham, North Carolina, United States of America

Department of Physics, Duke University, Durham, North Carolina, United States of America

Department of Computer Science, Duke University, Durham, North Carolina, United States of America

Department of Biochemistry, Duke University Medical Center, Durham, North Carolina, United States of America

INTRODUCTION

Intracellular electrodes fashioned from glass pipettes have been available for decades, but their fragility and high impedance motivates the search for novel probes with improved electrical and mechanical properties [1]–[18]. Indeed, diverse approaches have been used to improve recording electrodes, including new coating layers [1], [4], substrate-dependent platforms [2], [3], [5]–[10], new fabrication methods for metal electrodes [11], and incorporation of new materials [12]–[17]. Three requirements that any novel intracellular recording technology must meet are electrochemical stability, resistance to bio-fouling and mechanical compatibility with brain tissue. Carbon nanotubes (CNTs) display high electrical conductance, pronounced mechanical strength, and large surface area, which along with their demonstrated electrochemical and biological stability [18], indicate that they have the potential to meet these requirements. However, the difficulty of assembling CNTs has restricted their use to surface coating over a flat substrate or wires [1]–[3], limiting their previous application to in vivo extracellular recording from a cortex [1], stimulation on a separated muscle [2], or extracellular recording from isolated retinas [3].

An intracellular electrode made out of pure CNTs could exploit the attractive electromechanical properties of this material but requires a relatively long (~ 1 mm) insulated shaft to penetrate into brain tissue and an exposed tip of sub-micron diameter to impale and stably record from neuronal cell bodies, which are 5–50 μm in diameter in the vertebrate brain. With this goal in mind, we developed a procedure involving dielectrophoresis, annealing, insulation coating, and tip exposure to make a self-entangled, needle-shaped CNT probe suitable for obtaining intracellular recordings from vertebrate neurons.

MATERIALS AND METHODS

Dielectrophoresis

The self-entangled MWCNT probe was made by dielectrophoresis with an electrochemically sharpened tungsten wire (diameter 125 μm) and MWCNT dispersed in solution. The electrochemical etching process was described previously [19]. MWCNTs (outer diameter 8–15 nm, 95 wt%) were purchased from Cheap Tubes. The solution was prepared by 3 steps: mixing, sonication, and centrifugation. MWCNT 0.4 g, Polyvinylpyrrolidone (PVP, surfactant) 0.12 g, and deionized water (DIW) 40 ml were mixed and sonicated with a high-intensity probe type ultrasonic processor ([(30 sec maximum amplitude +10 sec pause) × 10 times] repeated 1 to 3 more times with ice cooling its container in between). Non-dispersed MWCNTs were precipitated by centrifuge (3,000 RPM, 20 minutes) and then discarded. The dielectrophoresis process [20], [21] used an electrochemically etched tungsten wire as the source electrode and a 25 mm diameter metal ring submerged beneath the surface of the MWCNT dispersed solution as a counter electrode. The sharp tip of the tungsten wire was placed to touch the solution in the middle of the counter electrode (see Figure 1). The tungsten wire and the counter electrode were then electrically connected to a power source, which supplied a sinusoidal 10 MHz signal, 40–80 V peak to peak amplitude. The tungsten wire was slowly pulled (40 μm/sec) out of the solution. The pulling speed was increased toward the end for growth termination at a desired length and to create a tapered end.

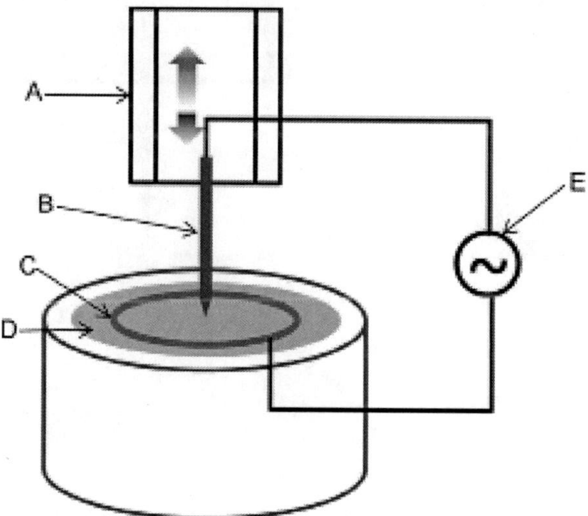

Figure 1. CNT fibril dielectrophoresis "pulling" stage assembled for this study.

(A) A motorized linear stage moves only in the vertical direction, pulling the tungsten wire out of the solution. (B) A electrochemically sharpened tungsten tip functioned as a source electrode. (C) A submerged metal ring functioned as a counter electrode. (D) CNT dispersed solution. (E) High-frequency AC power source.

Annealing

Using a micro-stage while monitoring the proximity with an optical microscope, the end of the CNT probe was placed to touch the top of a water droplet on a grounded surface (see Figure 2). DC voltage applied to the probe was ramped up to a threshold value around 80 V with a limited current by a 10 MΩ. When a threshold voltage was reached, a few microns of the probe tip got cut-off generating tiny water mist nearby.

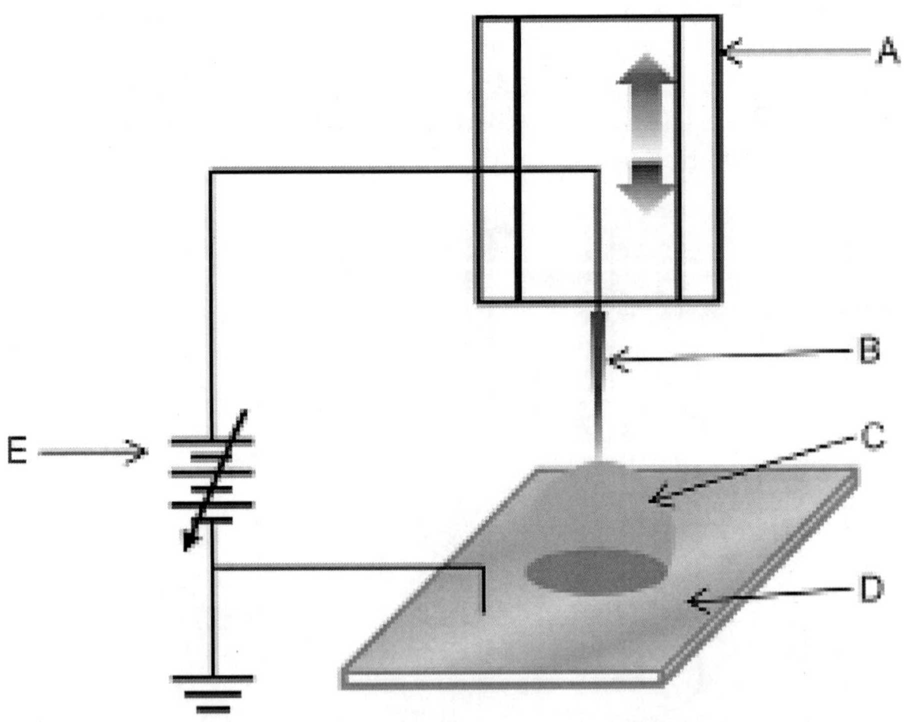

Figure 2. CNT fibril annealing setup.

(A) A motorized linear stage moving only in the vertical direction.

(B) A CNT probe. (C) Water droplet. (D) Gold plated surface for grounding. (E) Variable DC voltage source.

Parylene-C Coating and FIB (Focused Ion Beam) Tip Exposure

LPPVD (Low-Pressure Physical Vapor Deposition, Cookson Electronics PDS 2010 LABCOTER2) was used in-house for the coating. Around 250 nm thickness of Parylene-C was coated homogeneously in the deposition chamber where the CNT probes were hung downward in the middle of the chamber. For the tip exposure, FEI Quanta 200 3D FEG system was used. Ga+ beam was accelerated to 30 kV and its current was 1 nA for rough cutting and 100 pA for final trimming.

Cyclic Voltammetry and Impedance Spectroscopy

A three-electrode electrochemical cell was used with an impedance analyzer setup (ECI (Electrochemical Interface) and FRA (Frequency Response Analyzer) from Solatron Analytical). An AgCl/Ag pellet was used as the reference electrode, a metal mesh as the counter electrode, and 25 mM PBS (Phosphate buffered saline) is used to mimic the osmolarity of the brains. During these measurements, controlling the contact area between the CNT probe and phosphate buffered saline (PBS) is challenging, because wetting area of the solution on the CNR probe is not linearly proportional to dipping depth. Therefore different dipping depth into the PBS solution gives a range of measurement for any sample. Given that, both CV and EIS gave reliably repeatable results for each single sample (see Appendix S1 and Figure S1 for further details).

Intracellular Recording

Mice (Mus musculus, 34 days post natal, VGAT-ChR2-YFP line 8 (The Jackson Laboratory Stock Number: 014548, VGAT stands for vesicular glutamic acid transporter, ChR2 for channelrhodopsin

2, and YFP for Yellow Fluorescent Protein.) were used for the intracellular recording, in accordance with a protocol approved by the Duke University Institutional Animal Care and Use Committee under protocol A292-11-11. After induction of inhalation anesthesia (Isoflurane), the mouse was decapitated, and the brain was removed rapidly and placed in oxygenated ice-cold Ca-free artificial CSF (ACSF). Coronal brain slices were cut at 400 μm thickness and transferred to a holding chamber (room temperature) for 2–4 h. Individual slices were transferred to an interface-type chamber (30°C; Medical Systems) for intracellular recordings. The ACSF consisted of (in mM) 119 NaCl, 2.5 KCl, 1.3 MgC12, 2.5 CaC12, 1 NaH2PO4, 26.4 NaHCO3, and 11 glucose, equilibrated with 95% O2/5% CO2. Intracellular potentials were passed through a voltage follower (HS-2A ×0.1 L, Axon Instruments, leakage current 1 pA), low-pass filtered at 10 kHz, then digitized at 10 kHz. Targeted brain areas were cortex layer II/III or V. For an electrical stimulation, a concentric bipolar electrode (which looks like a blunt needle) was placed near recording site on the brain slice, and injected positive or negative step function current (typically 10 μA) briefly (1–10 ms). For the optical stimulation, an optical fiber was pointed directly above the recording site to deliver blue laser light (wavelength = 473 nm).

Extracellular Recording

All experiments were performed in accordance with protocol approved by the Duke University Institutional Animal Care and Use Committee under protocol A292-11-11. Male mice (Mus musculus, ~30 days old) were anesthetized with ketamine (46 mg/kg) and xylazine (24 mg/kg) and supplemented with isoflurane (0.5–1.5%) in oxygen. Core temperature was monitored with a rectal probe (BAT-12, Physitemp Instruments) and maintained at 36±1°C with a heating pad. A midline incision of the scalp was performed, and the skin was retracted. A small metal pin was attached to the cranium with dental cement, and the animal was fixed in a custom-made stereotaxic stage. A small craniotomy (~200 μm2) was performed over somatosensory cortex, and the dura overlying the brain was resected. The cortical surface was kept continuously moist. The electrode was lowered into the brain 100 - 1,000 μm with a hydraulic micromanipulator (SD Instruments). Signals were amplified 1000 times and band pass filtered (100 Hz –10 kHz) with a differential amplifier (A-M Systems) before being recorded with Spike 2 (Cambridge Electronic Design) at a sampling rate of 10 kHz.

RESULTS

The tip of an electrochemically sharpened tungsten wire served as the foundation from which a pure CNT probe was elaborated. We optimized the process to draw untapered probes of ~ 1 mm length and 5–10 μm diameter from a multi-walled CNT (MWCNT) solution. In the dielectrophoresis process, important parameters were: applied potential, speed of pulling, and concentration of dispersed CNTs in the dielectrophoresis solution. We found the diameter of CNT probes was proportional to amplitude of the applied potential and concentration of CNTs in the solution, and inversely to the speed of pulling. The combination of parameters was optimized for homogeneity of diameter along the length of the CNT probe. Finally, the tapering toward the last 100 μm was achieved by increased pulling speed. An annealing step stabilized the electrochemical performance of the CNT probe and improved its mechanical stiffness, but did not significantly alter its impedance characteristics. A 300 nm thick Parylene-C coating was applied along the length of the probe's shaft (Figure 3D) and a focused ion beam (FIB) was used to expose and shape the conductive CNT probe tip in an angled shape so that it resembled the blade of a flat-head screw driver (Figure 3E,F).

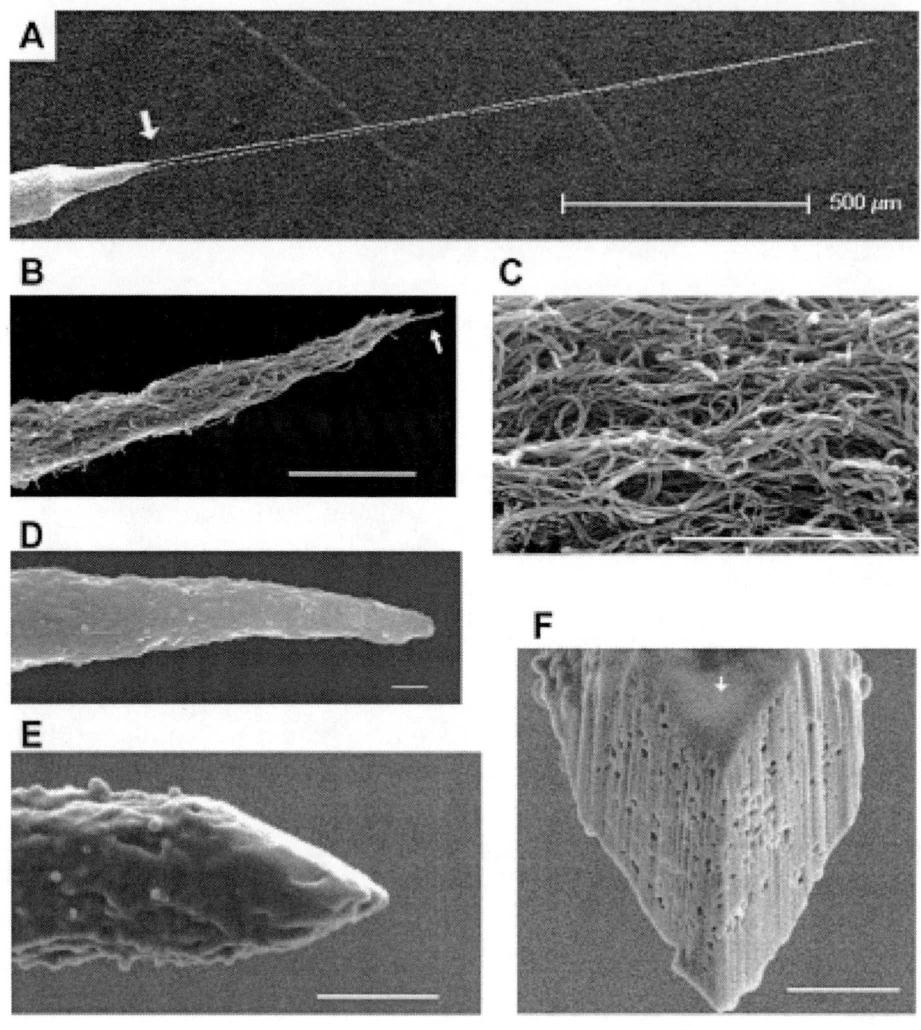

Figure 3. SEM images of CNT probes.

(A) Low magnification view of a CNT probe. Tungsten wire extends from lower left to the white arrow; CNT probe extends from the arrow for 1.5 mm to upper right. (B) Tip of a CNT probe after dielectrophoresis. The tip tapers down to a single CNT (white arrow). (C) CNTs in the probe show a clearly self-entangled morphology. (D) A CNT probe tip after coating with 300 nm Parylene-C, which homogeneously covers the entire probe. (E) An exposed CNT probe tip after FIB cutting. Two FIB cutting planes are perpendicular to the picture, crossing at the end of the probe. (F) An angled view (at 40° with respect to the electron beam in the SEM) of exposed CNT probe tip shows the two cut planes. The FIB cutting did not damage nearby insulation coating, which is clearly visible (white arrow). Scale bars in (B)- (F): 1 μm.

The electrochemical characteristics of the CNT probes (n = 5) were measured by cyclic voltammetry (CV) and electrochemical impedance spectroscopy (EIS). As Zheng et al. described extensively [22], a classical simple interface model is not applicable for an ionic liquid and CNT electrode interface. However, we can obtain the following qualitative information and comparisons from the measurements. There are no oxidation or reduction peaks in CV (Figure 4A) within the applied potentials ranging from −1 to +1 V, which indicates sufficient electrochemical inertness of the CNT probe. From the EIS measurement (Figure 4B), we can deduce the quality of insulation on the CNT probes and the recording signal quality. A properly applied insulation coating will increase the impedance by reducing contact area significantly, yet relatively low impedance (<100 MΩ) over typical operating frequencies (1 to 100 kHz) is desired, since it can improve signal quality. These data, coupled with SEM inspection of the CNT probe, imply that the increased impedance measured after Parylene-C coating and FIB cutting of the probe, is a direct result of our successfully reducing the contact area. Finally, impedance measurements revealed that the impedance of CNT probes (n = 5) were an order of magnitude lower than conventional glass electrodes (n = 3) over a frequency range of 1 to 100 kHz. These electrochemical features have two practical implications; (1) the lower impedance of the CNT probes can yield high signal to noise recordings. (2) the relatively large capacitance of the CNT probe (Figure 4A) is difficult to compensate during current injection state. Therefore, we used the CNT probes for passive voltage measurements only.

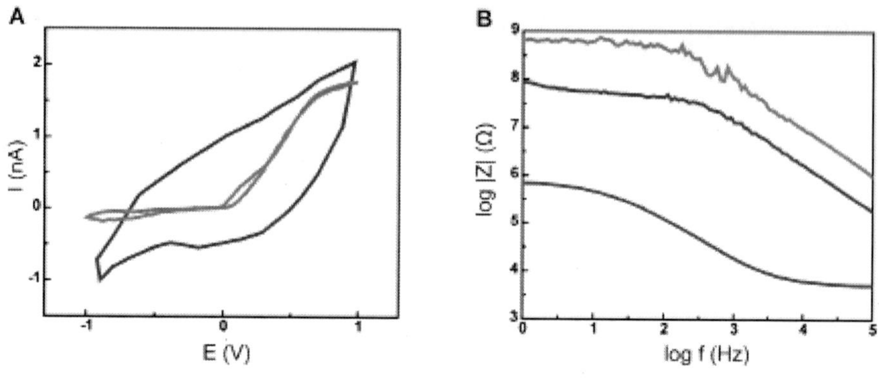

Figure 4. Impedance characterization of CNT probes.

(A) Cyclic Voltammetry (CV) shows the CNT probes (with the insulation and FIB cut) have significantly larger charge transfer (blue line), compared to the sharp intracellular-recording glass pipette (gray line). (B) Electrochemical impedance spectroscopy (EIS) shows significantly lower impedance of the CNT probe (blue line) compared to the glass pipette (gray line) over the frequency range of 1 to 100 kHz. Successful insulation-coating plus FIB-cutting reduces contact area between the CNT probe and solution by a few orders of magnitude, which explains the difference between the curves measured before coating (red line) and after FIB cutting (blue line). doi:10.1371/journal.pone.0065715.g004

To begin to explore the suitability of the CNT probes for neural recording, we first attempted to obtain intracellular recordings from cortical neurons in brain slices prepared from transgenic mice in which the light-activated ion channel channelrhodopsin 2 (ChR2) [23]–[25] was placed under the promoter for the vesicular glutamic acid transporter (VGAT), thus restricting expression of ChR2 to GABA-releasing inhibitory neurons. When a ChR2 protein absorbs blue light, it forms a nonspecific cation channel. Therefore, blue light stimulation brings cations into ChR2-expressing neurons that depolarize the cell and lead to firing of action potential(s). Focal illumination of the brain slice in the region of the recording site was then used to activate inhibitory synapses on an impaled cell, while a concentric bipolar stimulation electrode placed adjacent to the recording site to electrically stimulate a population of neurons which will drive a mixture of excitatory and inhibitory synaptic inputs. Using this approach, we were able to make intracellular recordings from a total of four cells using three different CNT probes (out of 6 CNT probes), with a mean recording time of 3 minutes (198.1±41.7 seconds, Figure 5 and Figure S2). Successful penetration was accompanied by a sharp drop in membrane potential (−57±4.62 mV), and enabled the detection of spontaneous synaptic potentials and action potentials (61.53±7.92 mV, Time course of action potentials is shown in Figure S3). Focal illumination with 473 nm light pulses (2 ms) delivered via a 200 μm (core diameter) micron fiber optic evoked short latency (4.94±0.84 ms) hyperpolarizing membrane potential responses (11.86±1.43 mV) in the impaled cell, consistent with ChR2-mediated activation of inhibitory inputs. Electrical stimulation evoked short latency depolarizing responses accompanied by action potentials or biphasic responses consisting of depolarizing and hyperpolarizing components, consistent with electrical activation of excitatory and inhibitory inputs. These observations indicate that the CNT probe developed here is suitable for intracellular recordings of membrane potential from cortical neurons.

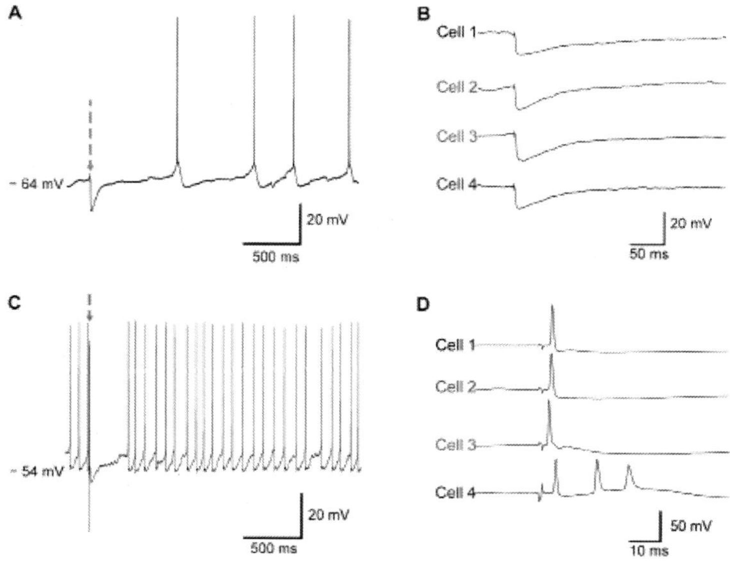

Figure 5. In vitro brain slice intracellular recording from mouse, cortical neurons.

(A) Light-evoked inhibitory post synaptic potential (IPSP) recorded with a CNT probe from a neuron in cortical brain slice prepared from a VGAT-CHR2 mouse. A brief light pulse was applied at the time marked by the arrow (Vm = −64 mV). (B) Expanded membrane potential records for CHR2 evoked IPSPs collected from four different cortical neurons. (C) Electrically-evoked excitatory postsynaptic potential (EPSP) from a cortical neuron in a mouse cortical brain slice (Vm = −54 mV). (D) Expanded membrane potential records of electrically-evoked EPSPs collected from four different cortical neurons.

We also explored whether the CNT probe could be used to record neural activity in the somatosensory cortex of the anesthetized mouse (n = 2). When advanced into the cortex, all seven CNT probe samples were able to detect single and/or multi-unit extracellular activity, in some cases revealing extremely large amplitude single unit events with large signal to noise ratio (S/N = 9 − 34, Figure 6). Although intracellular recording configurations were not achieved in these preliminary experiments, the successful intracellular recordings made in brain slices suggest that such recordings are feasible using CNT probes in vivo. An additional interesting observation was elastic deformation of the CNT probes: when a small amount of off-axis force was applied to the CNT probes brain tissue, the shaft of the probe bent elastically without snapping. Because mechanical resilience of a neural electrode is crucial for in vivo intracellular recordings, the elastic feature of the CNT probe could offer advantages over conventional glass electrodes for such an application.

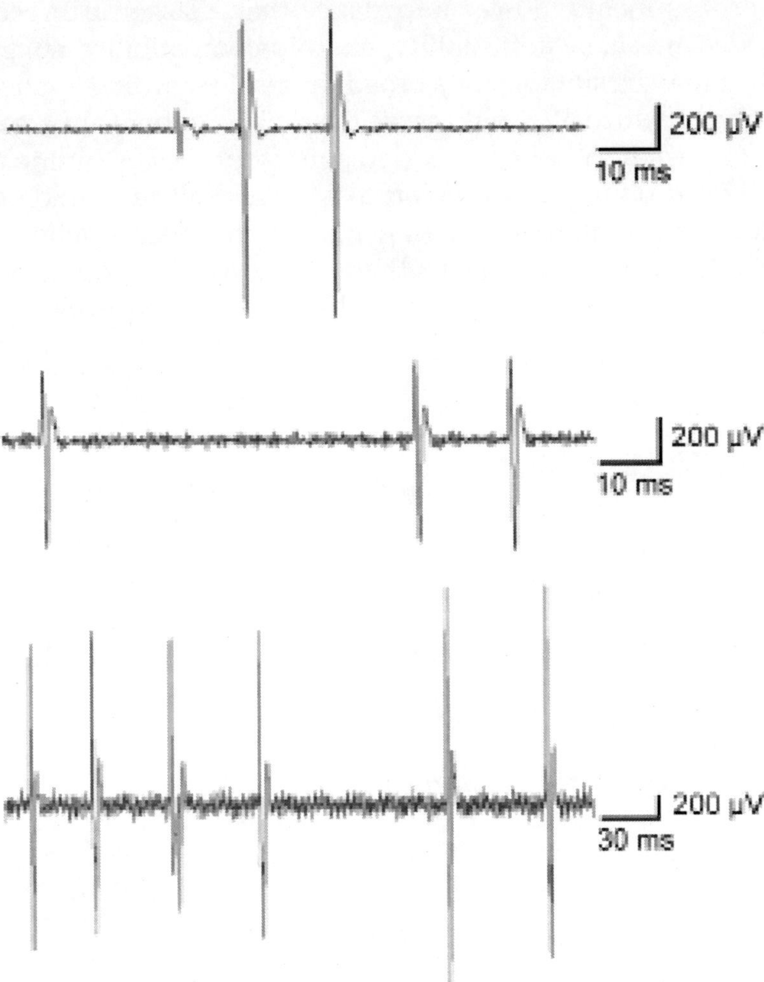

Figure 6. In vivo extracellular recordings from mouse brain somatosensory cortex.

Three different extracellular recordings show well-isolated single unit activity.

DISCUSSION

Here we developed a novel CNT probe suitable for intracellular and extracellular recordings from vertebrate neurons. The advantages of

CNT probes include lower impedance than glass sharp recording electrodes, mechanical flexibility, and bio-compatibility, suggesting that they may be suitable for a broad range of recording applications in vivo and in vitro. We could reuse the probes without any treatment as long as we didn't physically damage the probes by hitting a hard surface or applying excessive off-axis force. Full realization of this potential will require improving insulation layers, implementing capacitance compensation methods, and further refinement of the probe geometry. Nonetheless, the CNT probe developed here presents a new and simple device for use as an interface with neural tissue.

ACKNOWLEDGMENTS

We thank Dr. Igor Paprotny for initial discussions, Professors Warren Grill and Terrence G. Oas for lending us the impedance measurement setup and high frequency power equipment, Mr. Anders Nelson for help with extracellular recording, and Dr. Ka C. Wong for help with the FIB.

AUTHOR CONTRIBUTIONS

Conceived and designed the experiments: IY GF RM BRD. Performed the experiments: IY KH. Analyzed the data: IY KH IB RM BRD. Contributed reagents/materials/analysis tools: IY KH IB GF. Wrote the paper: IY RM BRD.

REFERENCES

1. Keefer EW, Botterman BR, Romero MI, Rossi AF, Gross GW (2008) Carbon nanotube coating improves neuronal recordings. Nat Nanotechnol 3: 434–439. doi: 10.1038/nnano.2008.174
2. de Asis ED Jr, Leung J, Wood S, Nguyen CV (2010) Empirical study of unipolar and bipolar configurations using high resolution single multi-walled carbon nanotube electrodes for electrophysiological probing of electrically excitable cells. Nanotechnology 21: 125101. doi: 10.1088/0957-4484/21/12/125101
3. Shoval A, Adams C, David-Pur M, Shein M, Hanein Y, et al.. (2009) Carbon nanotube electrodes for effective interfacing with retinal tissue. Frontiers in Neuroengineering 2.

4. Ensell G, Banks DJ, Ewins DJ, Balachandran W, Richards PR (1996) Silicon-based microelectrodes for neurophysiology fabricated using a gold metallization/nitride passivation system. Journal of Microelectromechanical Systems 5: 117–121. doi: 10.1109/84.506199
5. Normann RA, Maynard EM, Rousche PJ, Warren DJ (1999) A neural interface for a cortical vision prosthesis. Vision Research 39: 2577–2587. doi: 10.1016/S0042-6989(99)00040-1
6. Wise KD, Anderson DJ, Hetke JF, Kipke DR, Najafi K (2004) Wireless implantable microsystems: high-density electronic interfaces to the nervous system. Proceedings of the IEEE 92: 76–97. doi: 10.1109/jproc.2003.820544
7. Musallam S, Bak MJ, Troyk PR, Andersen RA (2007) A floating metal microelectrode array for chronic implantation. Journal of Neuroscience Methods 160: 122–127. doi: 10.1016/j.jneumeth.2006.09.005
8. Duan X, Gao R, Xie P, Cohen-Karni T, Qing Q, et al. (2012) Intracellular recordings of action potentials by an extracellular nanoscale field-effect transistor. Nat Nanotechnol 7: 174–179. doi: 10.1038/nnano.2011.223
9. Robinson JT, Jorgolli M, Shalek AK, Yoon MH, Gertner RS, et al. (2012) Vertical nanowire electrode arrays as a scalable platform for intracellular interfacing to neuronal circuits. Nat Nanotechnol 7: 180–184. doi: 10.1038/nnano.2011.249
10. Xie C, Lin Z, Hanson L, Cui Y, Cui B (2012) Intracellular recording of action potentials by nanopillar electroporation. Nat Nanotechnol 7: 185–190. doi: 10.1038/nnano.2012.8
11. Angle MR, Schaefer AT (2012) Neuronal Recordings with Solid-Conductor Intracellular Nanoelectrodes (SCINEs). PLoS One 7: e43194. doi: 10.1371/journal.pone.0043194
12. Yeh SR, Chen YC, Su HC, Yew TR, Kao HH, et al. (2009) Interfacing neurons both extracellularly and intracellularly using carbon-nanotube probes with long-term endurance. Langmuir 25: 7718–7724. doi: 10.1021/la900264x
13. Schrlau MG, Dun NJ, Bau HH (2009) Cell electrophysiology with carbon nanopipettes. ACS Nano 3: 563–568. doi: 10.1021/nn800851d
14. Moxon KA, Leiser SC, Gerhardt GA, Barbee KA, Chapin JK (2004) Ceramic-based multisite electrode arrays for chronic single-neuron recording. IEEE Transactions on Biomedical Engineering 51: 647–656. doi: 10.1109/TBME.2003.821037
15. Pellinen DS, Moon T, Vetter RJ, Miriani R, Kipke DR (2005) Multifunctional flexible parylene-based intracortical microelectrodes. Conf Proc IEEE Eng Med Biol Soc 7: 5272–5275. doi: 10.1109/IEMBS.2005.1615669
16. Jensen W, Yoshida K, Hofmann UG (2006) In-vivo implant mechanics of flexible, silicon-based ACREO microelectrode arrays in rat cerebral cortex. IEEE Transactions on Biomedical Engineering 53: 934–940. doi: 10.1109/TBME.2006.872824
17. Cheung KC, Renaud P, Tanila H, Djupsund K (2007) Flexible polyimide microelectrode array for in vivo recordings and current source density

analysis. Biosensors & Bioelectronics 22: 1783–1790. doi: 10.1016/j.bios.2006.08.035

18. Voge CM, Stegemann JP (2011) Carbon nanotubes in neural interfacing applications. J Neural Eng 8: 011001. doi: 10.1088/1741-2560/8/1/011001
19. Borzenets IV, Yoon I, Prior MM, Donald BR, Mooney RD, et al. (2012) Ultra-sharp metal and nanotube-based probes for applications in scanning microscopy and neural recording. J Appl Phys 111: 74703–747036. doi: 10.1063/1.3702802
20. Ma J, Tang J, Zhang H, Shinya N, Qin LC (2009) Ultrathin carbon nanotube fibrils of high electrochemical capacitance. ACS Nano 3: 3679–3683. doi: 10.1021/nn900787h
21. Tang J, Gao B, Geng H, Velev OD, Qin LC, et al. (2003) Assembly of 1D Nanostructures into Sub-micrometer Diameter Fibrils with Controlled and Variable Length by Dielectrophoresis. Advanced Materials 15: 1352–1355. doi: 10.1002/adma.200305086
22. Zheng JP, Goonetilleke PC, Pettit CM, Roy D (2010) Probing the electrochemical double layer of an ionic liquid using voltammetry and impedance spectroscopy: A comparative study of carbon nanotube and glassy carbon electrodes in [EMIM]+[EtSO4]−. Talanta 81: 1045–1055. doi: 10.1016/j.talanta.2010.01.059
23. Boyden ES, Zhang F, Bamberg E, Nagel G, Deisseroth K (2005) Millisecond-timescale, genetically targeted optical control of neural activity. Nat Neurosci 8: 1263–1268. doi: 10.1038/nn1525
24. Nagel G, Brauner M, Liewald JF, Adeishvili N, Bamberg E, et al. (2005) Light activation of channelrhodopsin-2 in excitable cells of Caenorhabditis elegans triggers rapid behavioral responses. Curr Biol 15: 2279–2284. doi: 10.1016/j.cub.2005.11.032
25. Nagel G, Szellas T, Huhn W, Kateriya S, Adeishvili N, et al. (2003) Channelrhodopsin-2, a directly light-gated cation-selective membrane channel. Proc Natl Acad Sci U S A 100: 13940–13945. doi: 10.1073/pnas.1936192100

Chapter 7

NOVEL NANOCOMPOSITES FROM SPIDER SILK–SILICA FUSION (CHIMERIC) PROTEINS

Cheryl Wong Po Foo [*], Siddharth V. Patwardhan [†], David J. Belton [†], Brandon Kitchel [*], Daphne Anastasiades [*], Jia Huang [*], Rajesh R. Naik [‡], Carole C. Perry[†,§], and David L. Kaplan [*]

*Departments of Biomedical Engineering, Chemistry, and Chemical and Biological Engineering, Bioengineering and Biotechnology Center, Tufts University, Medford, MA 02155;

†Biomolecular and Materials Interface Research Group, School of Biomedical and Natural Sciences, Nottingham Trent University, Nottingham NG11 8NS, United Kingdom; and

‡Materials and Manufacturing Directorate, Air Force Research Laboratory, 3005 Hobson Way, Wright-Patterson Air Force Base, OH 45433-7702

Edited by Charles R. Cantor, Sequenom, Inc., San Diego, CA, and approved May 8, 2006 (received for review February 10, 2006).

ABSTRACT

Silica skeletal architectures in diatoms are characterized by remarkable morphological and nanostructural details. Silk proteins from spiders and silkworms form strong and intricate self-assembling fibrous biomaterials in nature. We combined the features of silk with biosilica through the design, synthesis, and characterization of a novel family of chimeric proteins for subsequent use in model materials forming reactions. The domains from the major ampullate spidroin 1 (MaSp1) protein of Nephila clavipes spider dragline silk provide control over structural and morphological details because it can be self-assembled through diverse processing methods including film casting and fiber electrospinning. Biosilica nanostructures in diatoms are formed in aqueous ambient conditions at neutral pH and low temperatures. The R5 peptide derived from the silaffin protein of Cylindrotheca fusiformis induces and regulates silica precipitation in the chimeric protein designs under similar ambient conditions. Whereas mineralization reactions performed in the presence of R5 peptide alone form silica particles with a size distribution of 0.5–10 μm in diameter, reactions performed in the presence of the new fusion proteins generate nanocomposite materials containing silica particles with a narrower size distribution of 0.5–2 μm in diameter. Furthermore, we demonstrate that composite morphology and structure could be regulated by controlling processing conditions to produce films and fibers. These results suggest that the chimeric protein provides new options for processing and control over silica particle sizes, important benefits for biomedical and specialty materials, particularly in light of the all aqueous processing and the nanocomposite features of these new materials.

Complex mineralized composite systems in nature provide rich ground for insight into mechanisms of biomineralization and novel materials designs (1–4). Some of the more common sources of inspiration include seashells, insect exoskeletons, extracellular matrices involved in bone and other hard tissues, and biosilica skeletons. The formation of natural inorganic–organic composites is a multistep process, including the assembly of the extracellular matrix, the selective transportation of inorganic ions to discrete organized compartments with subsequent mineral nucleation, and growth delineated by preorganized cellular compartments. Silica skeletons found in nature are based on nanoscale composites wherein the organic components, usually proteins, are functional parts of the skeletal structures while also serving as silica-forming components (5, 6). As a result, materials' toughness is improved, strength is retained, and fine morphological control is achieved, all hallmark attributes of biological composites.

Silica is widespread in biological systems and serves different functions, including support and protection in single-celled organisms, such as diatoms through to higher plants and animals (7, 8). The remarkable morphological control *in vivo* that generates intricate patterns at small-length scales is species-specific and has attracted a great deal of interest in recent years because such features exceed the capabilities of present-day synthetic and technological approaches to materials engineering *in vitro*. In nature, the biosynthesis of biosilica from "silicon" *in vivo* occurs under mild ambient physiological conditions, around neutral pH and low temperatures of ≈4–40°C, and is facilitated by various biomolecules (5, 6, 9). Such conditions are in stark contrast to geochemical and industrial syntheses of silica *in vitro*, typically accomplished under much harsher conditions of higher temperatures and extremes of pH. The controlled formation of bioinspired silica structures with a range of proteins, peptides, and synthetic additives under various physical reaction environments has been reported (10). Various proteins have been isolated from biosilicas, in particular, siliceous frustules of a few selected diatoms of which some low-molecular-weight proteins, known as silaffins, and some even lower molecular weight compounds, termed polyamines, have also been suggested to play a crucial role in silica formation (11). *In vitro* studies of silica formation have also been performed by using synthetic variants of the R5 peptide that derives from the repeating motif found in silaffin proteins (11–16). Even though the lysine and serine groups are not posttranslationally modified in R5 peptide (unlike natural silaffins), this 19-aa unit is found to promote and regulate silica formation at neutral pH (under conditions that would not have been expected to generate silica). Specifically, the R5 peptide has been used to obtain spherical silica nanostructures by using various precursors (12, 14,15) as well as structures with different morphologies, including arch shapes and elongated fibers (13).

In the present article we describe a novel biomimetic nanocomposite approach to synthesize silica composites using fusion (chimeric) proteins. Fusion proteins have found applications in a wide spectrum of areas such as the biomedical field [including immunology, cancer research, and drug delivery (17–20)] and materials science [self-assembled materials (e.g., gels), quantum dot bioconjugates, sensors, and inorganic materials synthesis (21–27)]. Here we describe new silica-based nanocomposites formed from bioengineered fusion proteins that consist of two components (Fig. 1 *A*), and we propose a model for silk protein assembly into films and fibers and silica deposition onto the materials during the mineralization reactions (Fig. 1 *B*). One part of the fusion protein is the R5 peptide, known for precipitating silica as previously described. The second part of the fusion protein is a self-assembling domain based on the consensus repeat in the major ampullate spidroin protein 1 (MaSp1) protein of *Nephila clavipes* spider dragline silk, known for the formation of highly stable (β-sheet) secondary structures with impressive mechanical properties. Importantly, in these designs we exploit two critical lessons in materials science and engineering from nature: (*i*) nanoscale structural protein materials are used to optimize mechanical function and materials stability, and (*ii*) improved materials properties are gained through the control of nanoscale organic–inorganic interfaces and composite structural features.

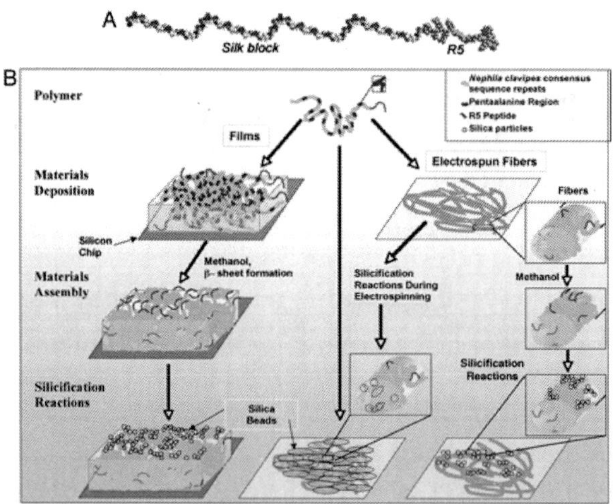

Figure 1. Schematic representation of the design of fusion proteins and their use in controlled silica nanocomposite formation. (*A*) Scheme of chimeric design with two functional domains: silk and R5. (*B*) Model of spider silk protein processing into films and fibers and silicification reactions on the assembled materials.

Silks are intriguing biologically derived proteins that form into fibers with remarkable mechanical properties (28, 29). In addition, silks self-assemble readily into defined β-sheet structures. Peptide variants of silkworm fibroin silk and spider dragline silks, as well as native reprocessed and genetic variants of these silks, have been studied to elucidate the important sequence chemistry–assembly relationships. To this end, a range of material morphologies and properties can now be generated through control of solution conditions, concentration, and additives, such that electrospun fibers (30, 31), films (32), porous matrices (33, 34), and hydrogels (35) can be generated under controlled conditions from these silk proteins that otherwise form into fibers only during processing in nature. The remarkable materials properties of these proteins prompt interest in their functionalization for enhancement in properties. For example, we have reported the successful chemical decoration of silk-based biomaterials with cell binding domains (36) and with cytokines such as bone morphogenetic protein 2 to enhance bone tissue formation (37).

The fusion proteins used in this study were generated by using a genetically engineered variant of a synthetic spider silk gene with the R5-encoding gene (Fig. 2) by using cloning strategies previously described (38–40). The amino acid sequences of the two spider dragline silk fusion proteins with silicification-inducing domains, with (type 1) and without (type 2) a CRGD cell-binding motif, are shown in Fig. 2.

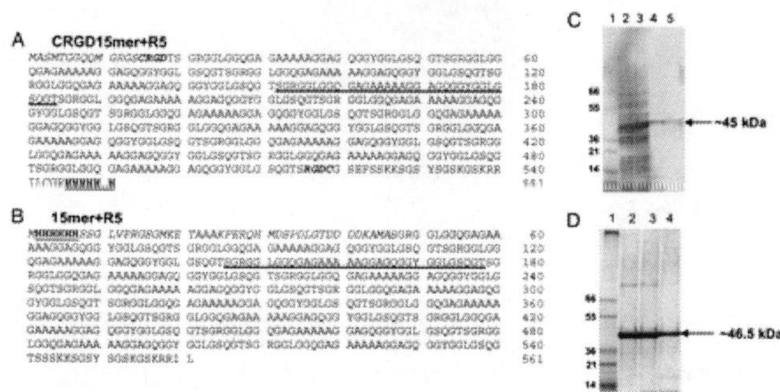

Figure 2. Expression and purification of spider silk fusion proteins. Amino acid sequences and gel electrophoresis of the bioengineered spider silk fusion proteins CRGD15mer+R5 (*A*) and 15mer+R5 (*B*). Underlined is the representa-

tive monomeric repeat unit selected and used in the design of the recombinant proteins based on the consensus sequence of spidroin 1 (Masp1) native sequence of *N. clavipes* (GenBank accession no. P19837). (*C*) Gel electrophoresis of 15mer+R5: lane 1, ladder; lane 2, flow-through; lane 3, wash 1; lane 4, elution 1; lane 5, elution 2. (*D*) Gel electrophoresis of CRGD15mer+R5: lanes 1, ladder; lane 2, elution 1; lane 3, elution 2; lane 4, elution 3 treated with 10 mM DTT.

Results and Discussion

When the R5 peptide was used alone at a ratio of one silicon to one amine from the side chains of R5 there was little difference in the rate of removal of soluble silicon species from solution (data not shown), suggesting little catalytic effect on the early stages of silicic acid polymerization. However, rapid precipitation of silica–peptide composites from the reaction media was observed (Fig. 3 *A*), consistent with previous studies (11, 13, 15). The precipitate formed showed that the silica formed is highly "dense" (surface area with and without R5 was 6.4 and 600 m^2/g, respectively). Upon calcination and removal of the organic phase from the silica–protein composite, the surface area increased to 520 m^2/g, implying occlusion of the peptide in the silica particles. The particles were found to be spheres of size ≈1 μm (Fig. 3 *B*). The role of R5 seems to be in aggregation and scaffolding rather than catalysis.

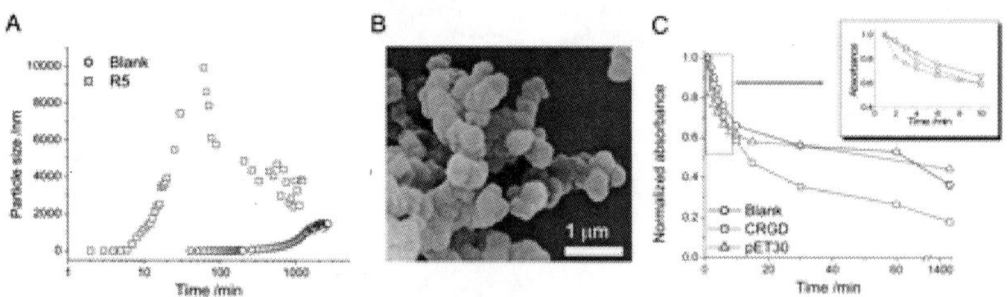

Figure 3. Effect of R5 peptide and fusion proteins on silicification. (*A*) Dynamic light scattering data for silica obtained from experiments performed in the presence of R5 alone. (*B*) Typical SEM of silica formed in the presence of 1 silicon:1 nitrogen-containing amino acid in R5 at pH 7.0 from aqueous solution. The ratio of silicon to nitrogen containing amino acid side chains in R5 was kept constant at 1:1 to allow comparison with our previous studies using simple amino acids, peptides, and small amines (44, 45). (*C*) Normalized

absorbance from soluble silicon released from a 30 mM aqueous solution of a silicon catecholato complex in the presence and absence of fusion proteins (silk+R5) over 24 h at pH adjusted to ≈6.8 as described previously (44, 45).

Furthermore, the silk+R5 fusion proteins were introduced into the silica polymerization experiments, data were obtained on kinetics of silicic acid polymerization, and morphology and porosity of silica precipitated. Even at low levels of fusion protein (≈135 silicon:1 R5 from fusion proteins, i.e., ≈22 silicon:1 amine groups from amino acid side chains of R5) there was an effect on the rate of removal of silicic acid from solution and on the nature of the silica phase formed (Fig. 3 *C*). Electron microscopy images of the precipitated silica showed networks of spheres of approximately >1 μm in diameter. These particles were similar in appearance to the silica produced in the presence of R5 alone where the silicon:amine from R5 was equal to one (Fig. 4). The elemental mapping of the samples (Fig. 4) revealed that the product contained silicon and oxygen arising from silica. The high amounts of carbon (Fig. 4 *B*) can be attributed to occluded protein. From the scanning electron microscopy (SEM) data (Fig. 4 *D*) it can be seen that, even without using any special assembly techniques, self-assembly of the silk chimera was evident in the presence of the silica structures. Thermal analysis was carried out on the samples, and the data suggest that ≈90% of the material was protein and the remaining 10% was silica. Nitrogen adsorption analysis of the composite samples indicated pore radii of the silica <10 Å and very low surface areas (≈10 m^2/g) when compared with blank samples (≈35 Å and ≈600 m^2/g, respectively).

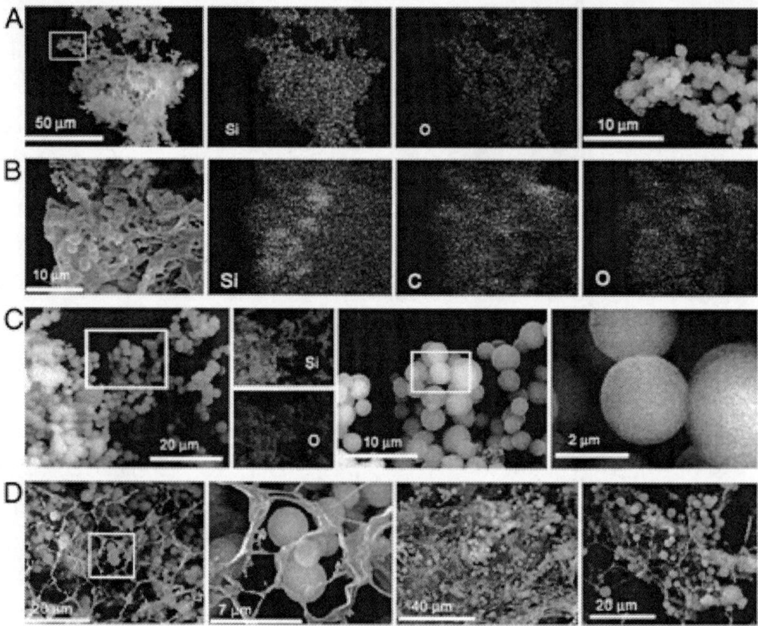

Figure 4. Morphological and elemental analyses of nanocomposites. SEM images of silica composite materials generated from an aqueous-based dipotassium silicon triscatecholate complex in the presence of chimera CRGD15mer+R5 (*A* and *B*) and chimera 15mer+R5 (Cand *D*). Areas highlighted by rectangles are presented at higher magnification in the same rows. In *A*–*C* elemental maps are shown for silicon (Si), oxygen (O), and/or carbon (C).

To exploit the self-assembling properties of silk in developing silk–silica nanocomposites, experiments were performed with tetramethoxysilane as the precursor. Four genetically engineered variants of the spider silk protein [two controls (one with and one without RGD but both without R5) and two chimeric versions of the controls with R5] were cast into films that were left untreated or were treated with methanol to induce a structural transition to β-sheets at the surface to decrease film solubility in aqueous buffer. Silicification reactions were performed on the films yielding spherical silica structures with diameters ranging from ≈0.5 to 2.0 μm only when the silica-precipitating domain, R5 peptide, was fused to the C terminus of the silk proteins (Fig. 5). The silk proteins that did not contain R5 (CRGD15mer and 15mer) did not yield significant changes in surface morphology of the films upon exposure to the silicification reactions.

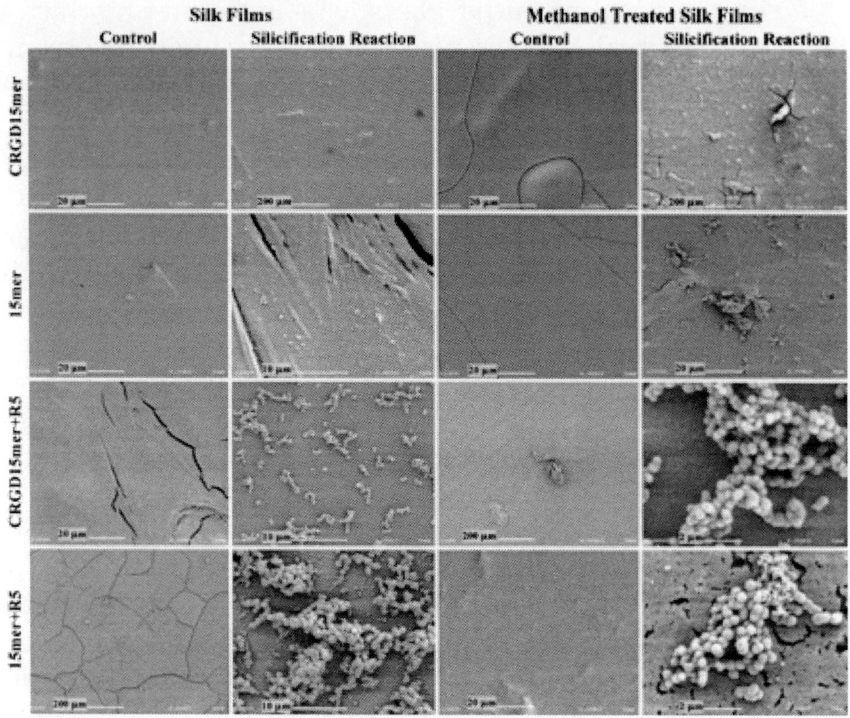

Figure 5. SEM images of untreated and methanol-treated silk films formed from four different genetically engineered silk proteins: CRGD15mer, 15mer, CRGD15mer+R5, and 15mer+R5. Images of the control films and the films that underwent silicification reactions are shown.

Fusion proteins were assembled into fibers by electrospinning (Fig. 6). SEM images of the electrospun fibers formed from the chimeric proteins (Fig. 6 *Ai*), and the morphological characteristics observed when the fibers were treated with methanol were similar to those we observed previously for electrospun silk fibroin with polyethylene oxide (31). Upon silicification on electrospun mats formed from the chimeric protein CRGD15mer+R5 without methanol treatment, similar spherical silica structures were observed as in the reactions on the cast films (Fig. 6 *Aii*). However, the dimensions of the silica spheres were slightly smaller, ranging from 200 to 400 nm (Fig. 6 *Aii*). When the electrospun fibers consisting of the chimera CRGD15mer+R5 were not treated with methanol, the fibers fused together on the surface (Fig. 6 *Aii*). Without the β-sheet inducing methanol treatment, the fibers are prone to partially solubilize on the surface, yielding fused fibers or a thin film on which the mineralization reaction takes place. However, upon treatment of the chimera CRGD15mer+R5 electrospun mats with methanol before silicification, the fibers fused to a much lesser extent at the surface as expected (31), compared with fibers that were not treated with methanol, and silica nanospheres were either sparingly observed or not observed at all (data not shown). When the chimera CRGD15mer+R5 was electrospun during the silica polymerization process (concurrent processing), silica deposition was induced in and on the fibers, and elliptically shaped silica particles fused to the fibers were observed (Fig. 6 *A iii--v*). The individual fibers appeared sticky and fused to each other to a greater extent, and silica deposition around the fibers provided a nonuniform coating instead of the usual heterogeneous distribution of silica nanospheres (Fig. 6 *A iii--v*). X-ray photoelectron spectroscopy analysis of the resulting fibers confirmed the presence of elemental silicon (Fig. 6 *B* and *C*). Thus, the concurrent processing approach, fiber spinning and silicification reactions, resulted in a different morphology of the silica in terms of location within the fibers and shape, compared with the silicification reactions conducted after electrospinning.

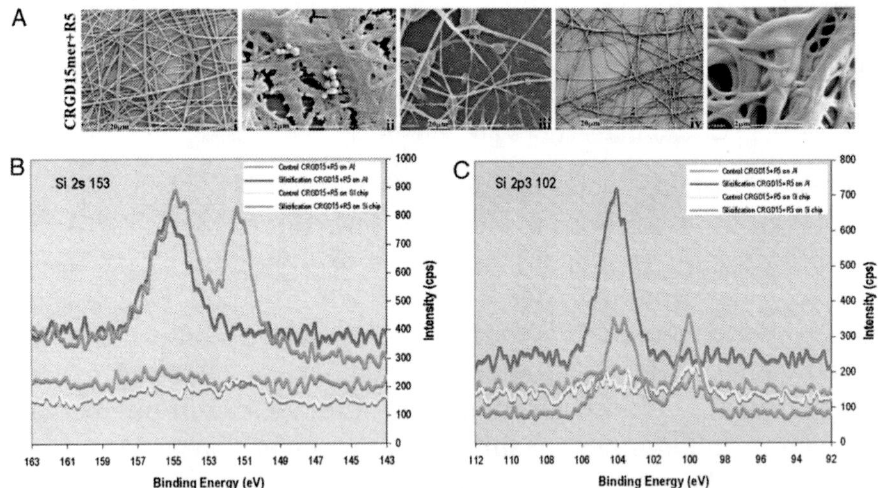

Figure 6. Morphological characterization and elemental analysis of silica deposition on eletrospun fibers. (*A*) SEM images of untreated and treated electrospun CRGD15mer+R5 silk fibers before, during, and after silicification reac-

tions. (*B*and *C*) X-ray photoelectron spectroscopy analysis of CRGD15mer+R5 and silicified CRGD15mer+R5 on Al foil and on silicon chip at the characteristic binding energies of 153 eV (*B*) and 102 eV (*C*) for electrons found in the 2s and 2p3 electron shells of the silicon atom, respectively

The design and use of novel chimeric fusion proteins containing silk and silica-forming domains in the synthesis of new silk–silica nanocomposites have been demonstrated. The properties of silk have been exploited to generate self-assembled composites in the form of films and fibers as examples to illustrate the diversity of processing options with this approach. Changes in processing conditions altered the size distribution as well as the morphology of the composites such that control of process details will provide control of the mineral phase and thus composite properties. The protein biomaterial self-assembling nanodomains used in these designs are genetically tailorable in terms of size, chemistry, and morphology, such that this approach offers a new platform for *in situ* silica formation with unprecedented control in composite materials design and properties. The silica-forming domain is also tailorable in terms of sequence chemistry to influence reaction kinetics and materials features. The silica-forming domains can also be replaced with alternative fusions and thereby be used to form other inorganic phases (e.g., hydroxyapatite, titanium dioxide, and germania); thus, the versatility and opportunities that can be explored with this chimeric biomimetic protein design strategy are expansive. This approach to nanoscale material composite systems engineering, we believe, will generate new families of biomaterials that can be either preassembled *in vitro* or organized (self-assembled) *in vivo*. The novelty and potential utility of the strategy described are further extended by the molecular-level connections between the organic (silk) and inorganic (silica) phases because of the chimeric design of the protein chains, as well as because of the all aqueous and ambient conditions under which materials formation is conducted. Specifically, the all-aqueous conditions offer future options to consider *in situ* reactions in tissue-compatible compartments *in vivo*; the domain sizes of the silica particles formed are in a size scale that makes them suitable candidates for *in vivo* needs as glassy biomaterials, compared with larger bulk silica or bioactive glass systems usually used as bioactive composites (41, 42). The ability to direct the silica-forming reactions to material interfaces is a plus in regulating materials morphology and thus function. This attribute is derived from the water-based processing and the resulting hydrophobic–hydrophilic partitioning of the hydrophobic silk protein component of the chimera and the hydrophilic R5 component.

MATERIALS AND METHODS

Design and Expression of Recombinant Spider Silk Protein

One repeat was designed and constructed for cloning by using synthetic oligonucleotides and then amplified by using PCR. The 15mer encoding the repeat was cloned through the transfer of cloned inserts between two shuttle vectors based on pUC19 and pCR-Script (39, 40) (Stratagene). The cloning of CRGD15mer recombinant protein was performed in a similar manner. Oligonucleotides for the R5 peptide were designed with EcoRI (gaattc) and NotI (gcggccgc) restriction sites at the 5′ and 3′ ends, respectively, and then ligated directly into the EcoRI and NotI restriction sites of pET-30a(+) vector (Novagen) next to the 15mer clone (Fig. 7, which is published as supporting information on the PNAS web site). The constructs pET-30a(+)-15mer+R5 and pET-21a(+)-CRGD15mer+R5 were transformed into the *Escherichia coli* host strain RY-3041, a mutant strain defective in the expression of SlyD protein, for protein expression (38, 43). The resulting proteins were finally purified under denaturing conditions using an Ni-NTA resin (Qiagen, Valencia, CA), which allows for the specific binding of the 6×His fusion tag at the C terminus of the proteins. The purified proteins were then identified by using SDS/PAGE (Invitrogen) (Fig. 2 *C* and *D*).

Preparation of Silica Samples

To establish baseline model reaction conditions for silica polymerization in the presence of R5, two different silica precursors were used, tetramethoxysilane and a silicon–catecholate complex (44, 45). Typically, for silica precipitation with or without the presence of proteins, 30 mM silicic acid was used, unless otherwise stated. Silica synthesis was carried out as described previously (11, 12, 14, 15, 44,

45). The reaction mixture was allowed to condense for a desired period while aliquots at known time were taken for molybdosilicate kinetic assay (44). Silica samples were isolated by centrifugation, washed, and lyophilized for SEM and nitrogen adsorption analyses. Nitrogen gas adsorption/desorption analysis was carried out by using a Quantachrome Nova3200e surface area and pore size analyzer. Surface areas were determined through the Brunauer–Emmett–Teller method (46), and pore radii were determined by the Barrett–Joiner–Halenda method (47) using the desorption branch of the isotherm. The entrapped organic material in silica was removed by calcination of samples at 650°C in air. Samples for SEM were prepared by dispersing the lyophilized powder on sample holders containing double-sided sticky carbon tape and gold coated under argon plasma.

Preparation of Silk Films

The lyophilized fusion proteins were dissolved in hexafluoroisopropanol at a concentration of 2.5% wt/vol at 4°C. A total of 100 μl of the protein–hexafluoroisopropanol solution was pipetted directly onto chemically inert silicon chips placed at the bottom of 24-well culture plates and allowed to air-dry. The silk films were then left untreated or were treated with a 90% vol/vol methanol/water mixture to induce β-sheet formation on the films surfaces and therefore prevent resolubilization of the films in aqueous solutions.

Silicification Reactions on Cast Films

A total of 200 μl of 100 mM phosphate buffer at pH 5.5 was added to cover the air-dried silk films formed on the silicon chips and were left to incubate for 30 min at room temperature. A total of 20 μl of 1 M tetramethoxysilane hydrolyzed in 1 mM hydrochloric acid was then added for the silicification reaction to take place, and the reaction was left to incubate for ≈10 min (37). The films were then washed with 18.2 MΩ water three times and left to dry overnight in the fume hood. The morphologies of the exposed surface of these processed silk films with silica deposition were then analyzed by using a LEO 982 Scanning Electron Microscope (Harvard University Center for Nanoscale Systems, Cambridge, MA).

Electrospinning of Silk Fibers and Silicification Reactions

Fibers of the recombinant spider silk proteins were electrospun directly onto silicon chips placed on the Al foil-covered receiving plate by using a 2% wt/vol solution of recombinant spider silk in hexafluoroisopropanol as previously described (31). The concentrated silk solution was infused at the rate of 0.01 ml/min at 15–20 kV. Some of the electrospun fibers were then treated with 90% vol/vol methanol in water to induce β-sheet formation by incubation for 10 min and were finally allowed to air-dry.

The same procedure as for the cast films was used to perform the silicification reactions on both methanol-treated and nonmethanol-treated electrospun fiber mats. Furthermore, silicification reactions were performed during electrospinning by injecting the silicic acid solution mixed with phosphate buffer at the same rate and time as the concentrated silk solution.

SEM was used to perform morphological characterization of the electrospun fibers, with and without methanol treatment and silicification reactions, by using a Leo 982 Field Emission Scanning Electron Microscope (Harvard University Center for Nanoscale Systems). Furthermore, elemental analysis of the unreacted and reacted samples was performed by using an x-ray photoelectron spectrometer equipped with an Al K_a radiation source and four available spot sizes ranging from 150 to 1,000 mm (Harvard University Center for Nanoscale Systems).

- §To whom correspondence may be addressed at:

Interdisciplinary Biomedical Research Centre, School of Biomedical and Natural Sciences, Nottingham Trent University, Clifton Lane, Nottingham NG11 8NS, United Kingdom . E-mail: carole.perry@ntu.ac.uk

- ¶To whom correspondence may be addressed at:

Department of Biomedical Engineering, Bioengineering and Biotechnology Center, Tufts University, 4 Colby Street, Medford, MA 02155 . E-mail: david.kaplan@tufts.edu

ACKNOWLEDGMENTS

We thank the National Institutes of Health (Grants EB00252 and EB003210), the U.S. Air Force Office of Scientific Research (Grants FA9550041 and F49620-03-1-0099), and the European Commission (SILIBIOTEC Project QLK3-CT-2002-01967) for their financial support.

AUTHOR CONTRIBUTIONS

C.W.P.F., S.V.P., J.H., C.C.P., and D.L.K. designed research; C.W.P.F., S.V.P., D.J.B., B.K., D.A., and C.C.P. performed research; R.R.N. contributed new reagents/analytic tools; C.W.P.F., S.V.P., D.J.B., and C.C.P. analyzed data; and C.W.P.F., S.V.P., C.C.P., and D.L.K. wrote the paper.

CONFLICT OF INTEREST STATEMENT:

No conflicts declared. This paper was submitted directly (Track II) to the PNAS office.

REFERENCES

1. Heuer A. H., Fink D. J., Laraia V. J., Arias J. L., Calvert P. D., Kendall K., Messing G. L., Blackwell J., Rieke P. C., Thompson D. H., et al. (1992) Science 255:1098–1105.
2. Boskey A. L. (1998) J. Cell. Biochem. Suppl. 30–31:83–91.
3. Li C., Vepari C., Jin H. J., Kim H. J., Kaplan D. L. (2006) Biomaterials 27:3115–3124.
4. Wilt F. H. (2005) Dev. Biol. 280.15–25.
5. Muller W. E. G. (2003) Silicon Biomineralization (Springer, Berlin).
6. Simpson T. L., Volcani B. E. (1981) Silicon and Siliceous Structures in Biological Systems(Springer, New York).
7. Wong Po Foo C., Huang J., Kaplan D. L. (2004) Trends Biotechnol. 22:577–585.
8. Perry C. C., Keeling-Tucker T. (2000) J. Biol. Inorg. Chem. 5:537–550.
9. Voronkov M. G. (1977) Silicon and Life (Zinatne, Vilnius, Lithuania).
10. Patwardhan S. V., Clarson S. J., Perry C. C. (2005) Chem. Commun. 9:1113–1121.
11. Sumper M., Kroger N. (2004) J. Mater. Chem. 14:2059–2065.

12. Brott L. L., Pikas D. J., Naik R. R., Kirkpatrick S. M., Tomlin D. W., Whitlock P. W., Clarson S. J., Stone M. O. (2001) Nature 413:291–293.
13. Naik R. R., Whitlock P. W., Rodriguez F., Brott L. L., Glawe D. D., Clarson S. J., Stone M. O. (2003) Chem. Commun. 238–239.
14. Patwardhan S. V. (2003) Ph.D. dissertation (Univ. of Cincinnati, Cincinnati).
15. Whitlock P. W. (2005) Ph.D. dissertation (Univ. of Cincinnati, Cincinnati).
16. Knecht M. R., Wright D. W. (2003) Chem. Commun. 3038–3039.
17. Albarran B., To R., Stayton P. S. (2005) Protein Eng. Des. Selection 18:147–152.
18. Asai T., Trinh R., Ng P. P., Penichet M. L., Wims L. A., Morrison S. L. (2005) Biomol. Eng.21:145–155.
19. Icke C., Schlott B., Ohlenschlager O., Hartmann M., Guhrs K. H., Glusa E.
20. (2002) Mol. Pharmacol. 62:203–209.
21. Park E., Starzyk R. M., McGrath J. P., Lee T., George J., Schutz A. J., Lynch P., Putney S. D. (1998) J. Drug Targeting 6:53–64.
22. Dubrovsky T., Tronin A., Dubrovskaya S., Guryev O., Nicolini C. (1996) Thin Solid Films285:698–702.
23. Frey W., Meyer D. E., Chilkoti A. (2003) Adv. Mater. 15:248–251.
24. Medintz I. L., Uyeda H. T., Goldman E. R., Mattoussi H. (2005) Nat. Mater. 4:435–446.
25. Padilla J. E., Colovos C., Yeates T. O. (2001) Proc. Natl. Acad Sci. U.S.A 98:2217–2221.
26. Paternolli C., Ghisellini P., Nicolini C. (2002) Mater. Sci. Eng. C. 22:155–159.
27. Sleytr U. B., Schuster B., Pum D. (2003) IEEE Eng. Med. Biol. Mag. 22:140–150.
28. Slocik J. M., Naik R. R., Stone M. O., Wright D. W. (2005) J. Mater. Chem. 15:749–753.
29. Cunniff P. M., Fossey S. A., Auerbach M. A., Song J. W. (1994) Silk Polymers (Am. Chem. Soc.Washington, DC), Vol. 544, pp 234–251.
30. Gosline J. M., Demont M. E., Denny M. W. (1986) Endeavour 10:37–43.
31. Wong Po Foo C., Bini E., Hensman J., Knight D. P., Lewis R. V., Kaplan D. L. (2006) Appl. Phys. A. 82:223–233.
32. Jin H. J., Fridrikh S. V., Rutledge G. C., Kaplan D. L. (2002) Biomacromolecules 3:1233–1239.
33. Jin H. J., Park J., Valluzzi R., Cebe P., Kaplan D. L. (2004) Biomacromolecules 5:711–717.
34. Kim U. J., Park J., Kim H. J., Wada M., Kaplan D. L. (2005) Biomaterials 26:2775–2785.
35. Nazarov R., Jin H. J., Kaplan D. L. (2004) Biomacromolecules 5:718–726.
36. Kim U. J., Park J. Y., Li C. M., Jin H. J., Valluzzi R., Kaplan D. L. (2004) Biomacromolecules5:786–792.
37. Sofia S., McCarthy M. B., Gronowicz G., Kaplan D. L. (2001) J. Biomed. Mater.

Res. 54:139–148.

38. Karageorgiou V., Meinel L., Hofmann S., Malhotra A., Volloch V., Kaplan D. (2004) J. Biomed. Mater. Res. 71A:528–537.

39. Huang J., Valluzzi R., Bini E., Vernaglia B., Kaplan D. L. (2003) J. Biol. Chem. 278:46117–46123.

40. Prince J. T., McGrath K. P., Digirolamo C. M., Kaplan D. L. (1995) Biochemistry 34:10879–10885.

41. Winkler S., Wilson D., Kaplan D. L. (2000) Biochemistry 39:12739–12746.

42. Boccaccini A. R., Notingher I., Maquet V., Jerome R. (2003) J. Mater. Sci. Mater. Med.14:443–450.

43. Wang M. (2003) Biomaterials 24:2133–2151.

44. Yan S. Z., Beeler J. A., Chen Y., Shelton R. K., Tang W. J. (2001) J. Biol. Chem.276:8500–8506.

45. Belton D., Paine G., Patwardhan S. V., Perry C. C. (2004) J. Mater. Chem. 14:2231–2241.

46. Belton D.,

47. Patwardhan S. V., Perry C. C. (2005) Chem. Commun. 3475–3477.

48. Brunauer S. , Emmett P. H. , Teller E. (1938) J. Am. Chem. Soc. 60:309–319.

49. Barrett E. P., Joyner L. G. , Halenda P. P. (1951) J. Am. Chem. Soc. 73:373–380.

Chapter 8

ON THE ELECTRICAL AND THERMAL CONDUCTIVITIES OF CAST A356/AL_2O_3 METAL MATRIX NANOCOMPOSITES

El-Sayed Youssef El-Kady, Tamer Samir Mahmoud, Ali Abdel-Aziz Ali

Mechanical Engineering Department, Faculty of Engineering, King Khalid University (KKU), Abha, Kingdom of Saudi Arabia.

Email: Eyelkady@yahoo.com

ABSTRACT

To assess the effect of the dispersion of Al2O3 nanoparticles into A356 Al alloy on both the electrical and thermal conductivities, A356/ Al_2O_3 metal matrix nanocomposites (MMNCs) were fabricated using a combination of rheocasting and squeeze casting techniques. Two different sizes of Al_2O_3 nanoparticles were dispersed into the A356 Al alloy, typically, 60 and 200 nm with volume fractions up to 5 vol%. The effect of the nanoparticles size and volume fraction on the electrical and thermal conductivities was evaluated. The results revealed that the A356 monolithic alloy exhibited better electrical and thermal conductivities than the MMNCs. Increasing the nanoparticles size and/or the volume fraction reduces both the thermal and electrical conductivities of the MMNCs. The maximum reduction percent in the thermal and electrical conductivities, according to the A356 monolithic alloy, were about 47% and 38%, respectively. Such percentages were exhibited by A356/Al2O3 MMNCs containing 5 vol% of nanoparticles having 60 and 200 nm, respectively.

INTRODUCTION

Metal matrix nanocomposites (MMNCs) reinforced by nanoscale particulates is one type of nanocomposite. MMNCs reinforced with nanoparticles are widely used in industry. Several applications of Al and Mg MMNCs are found in automotive and aerospace industry [1,2]. Conventional particulate reinforced metal matrix composites (MMCs) usually have a metallic phase (usually Al or Mg) and a ceramic reinforcement phase (most commonly SiC, Al_2O_3 and graphite particulates) composed of micron scale particles. Ingot metallurgy (I/M) (or casting techniques) and powder metallurgy (P/M) are common methods to produce conventional MMCs [2]. MMCs have higher yield strength, stiffness and lower coefficient of thermal expansion than those of the monolithic matrix alloy. Currently, there are several fabrication methods of MMNCs, including mechanical alloying, powder metallurgy, casting techniques, electrochemical deposition, friction stir processing, laser and sol-gel technology [1]. Each of these methods has its own advantages and disadvantages. Mechanical alloying, powered metallurgy and casting technique are the most commonly used techniques for producing bulk MMNCs [3]. With electrochemical deposition, laser technology and sol-gel fabrication methods, only thin films or layer of nanostructured MMNCs are formed.

The rheocasting (compocasting), as a semi-solid phase process, can produce good quality MMCs [4,5]. The technique has several advantages: It can be performed at temperatures lower than those conventionally employed in foundry practice during pouring; resulting in reduced thermochemical degradation of the reinforced surface, the material exhibits thixotropic behaviour typical of stircast alloys, and production can be carried out using conventional foundry methods. The preparation procedure for rheocast composites consists of the incorporation of the ceramic particles within very vigorously agitated semi-solid alloy slurry which can achieve more homogenous particles distribution as compared with a fully molten alloy. This is because of the presence of the solid phase in the viscous slurry that prevents the ceramic particles from settling and agglomerating. However, the composites produced by rheocasting suffer from the high porosity content, which has a deleterious effect on the mechanical properties. It has been reported that the porosity can be reduced by means of mechanical working such as extrusion or rolling on the solidified composites or by applying a pressure to the composite slurry during solidification [6,7].

However the mechanical and tribological characteristics of MMNCs were extensively studied [8,9], only limited data on the thermal and electrical behavior of such nanocomposites are available [10,11]. Until now the thermal and electrical behaviour of MMNCs were not sufficiently determined and published. Accordingly, the aim of the present work is to study the effect of the Al_2O_3 nanoparticles additives on both the electrical and the thermal conductivities of A356 Al alloy. The effect of the Al_2O_3 nanoparticles size and volume fraction of the aforementioned conductivities was evaluated. The A356/ Al_2O_3 MMNCs were fabricated a combination of rheocasting and squeeze casting techniques.

Experimental Procedures

The A356 Al-Si-Mg cast alloy was used as a matrix. The chemical composition of the A356 Al alloy is listed in **Table 1**. Nano-Al_2O_3 particulates were used as reinforcing agents. The Al2O3 nanoparticles have two different average sizes, typically, 200 and 60 nm. Several A356/ Al_2O_3 MMNCs were fabricated with different volume factions of nanoparticles up to 5 vol%.

The A356/Al_2O_3 MMNCs were prepared using a combination of rheocasting and squeeze casting techniques. Preparation of the composite alloy was carried out ac- cording to the following procedures: About 1 kg of the A356 Al alloy was melted at 680°C in a graphite crucible in an electrical resistance furnace. After complete melting and degassing by argon gas of the alloy, the alloy was allowed to cool to the semisolid temperature of 602°C. At such temperature the liquid/solid fraction was about 0.7. The liquid/solid ratio was determined using primary differential scanning calorimeter (DSC) experiments performed on the A356 alloy. A simple mechanical stirrer with three blades made from stainless steel coated with bentonite clay was introduced into the melt and stirring was started at approximately 1000 rpm. Before stirring the nanoparticles reinforcements after heating to 400°C for two hours were added inside the vortex formed due to stirring. After that, preheated Al_2O_3 nanoparticles were introduced into the matrix during the agitation. After completing the addition of Al_2O_3 nanoparticles, the agitation was stopped and the molten mixture was poured into preheated tool steel mould and immediately squeezed during solidification using a hydraulic press of 50 ton capacity for 5 minutes.

Table 1. The chemical composition (wt%) of the A356 alloy.

Si	Fe	Cu	Mn	Mg	Zn	Al
6.6	0.25	0.11	0.002	0.14	0.026	Bal.

Figure 1. A photograph of the apparatus used to measure the thermal conductivity of the nanocomposites.

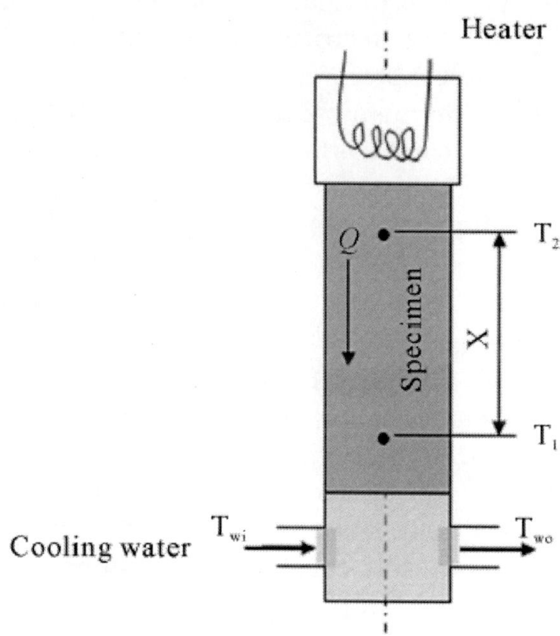

Figure 2. A schematic drawing shows the specimen, the positions of the thermocouples and the heating and cooling methods of the specimen.

The MMNCs were heat treated at T6 before testing. The MMNCs were solution treated at 540°C ± 1°C for three hours and then quenched in cold water. After cooling specimens were artificially aged at 160°C ± 1°C for 12 hours.

Samples from the fabricated composites were cut from the cast ingot for microstructural examinations. Specimens were ground under water on a rotating disc using SiC abrasive discs of increasing finesse up to 1200 grit. Then they were polished using 10 **μm** alumina paste and 3 **μm** diamond paste. Microstructural examinations were conducted using both optical and scanning electron microscopes (SEM). Microstructural examination was per- formed in the unetched condition. The porosity of the MMNCs was measured using the typical Archimedes (water displacement) method.

The electrical measurements of the A356 monolithic alloy and the A356/Al2O3 MMNCs were measured using SIGMASCOPE SMP10 apparatus made by Helmut Fischer GmbH + Co.KG, Germany. The apparatus can measure the electrical conductivity of all non-magnetic metals and even stainless steel, etc. Samples from the MMNCs and the monolithic alloy having a cylindrical shape of 20 mm diameter and 10 mm length were cut from the cast and polished from the two end faces before electrical conductivity measurements. The electrical conductivity measurements were performed according to ASTM E-1004. The apparatus measures the electrical conductivity using the eddy current method.

The thermal conductivity of the MMNCs was measured using the apparatus shown in **Figure 1**. The apparatus has been specially designed for the determination of thermal conductivity for both good conductors and thin specimens of insulants. The apparatus consists of a self- clamping specimen stack assembly with electrically heated source, calorimeter base, Dewar vessel enclosure to ensure negligible loss of heat, and constant head cooling water supply tank. A multipoint thermocouple switch is mounted on the MMNCs specimen and two mercury and glass thermometers are provided for water inlet and outlet temperature readings as shown in **Figure 2**.

Three NiCr/NiAl thermocouples are fitted and connections are provided for a suitable potentiometer instrument to give accurate temperature readings. The specimens have cylindrical shape of about 10 mm diameter and 20 mm length. The end faces of these specimens are very carefully prepared by grinding and lapping operations. Two small holes were drilled in each specimen for insertion of the thermocouples. Heat was applied to the top of the stack of specimens by contact with the thermostatically protected heater block, and is collected at the base of the stack by contact with a water cooled calorimeter. The heat transmitted through the specimens can thus be calculated from a measurement of water flow and temperature rise of the water. Heat losses by radiation and conduction from the heater block other than from contact with the specimen were neglected, since measurement is made of heat collected from the samples rather than heat delivered to them. A steady flow of cooling water is maintained at the end away from the heating block and it leaves at the end nearer to it. Thermometers *Twi* and *Two* are provided to measure the temperatures of the inlet and outlet water, respectively. Two holes of 1.5 mm diameter are drilled in the specimen rod and thermocouples are inserted in these holes to measure the temperatures *T*1 and *T*2 of the rod at these places. The temperatures of the two thermometers and the two thermocouples rise initially and ultimately become constant when the steady state is reached. The water coming out of the apparatus is collected in a beaker for a fixed time measured by a stopwatch and the flow rate of water is determined by timing the collection of 100 cm3 per second sample of water. The cross-section area of the rod is calculated by measuring its diameter and the distance between the holes in the rod was measured.

It is assumed that the cross-sectional area normal to the flow of heat transfer is constant and the periphery is insulated due to the use of Dewar vessel enclosure to ensure negligible loss of heat. The heat transfer through the specimen rod is governed by Fourier's Law:

$$Q = -kA\frac{dT}{dx} \qquad (1)$$

where Q is the rate of heat transfer in the x-direction, k is the thermal conductivity of the specimen material, A is the cross-sectional area of the specimen normal to the x-direction, and dT/dx is the temperature gradient in the x-direction. At steady state condition the temperature of the specimen does not change with time at any point and the heat conducted along the specimen must go into the flowing water. The heat taken by the water was calculated using the following equation:

$$Q = mc_p\left(T_{wo} - T_{wi}\right) \qquad (2)$$

where m is the mass flow rate of cooling water, Cp is the specific heat of water, and Two, Twi are the cooling water outlet and inlet temperatures, respectively. By applying a thermal balancing hence, the same rate of heat transfer Q is used in Equation (1) and (2). Thus,

$$k = \frac{mc_p\left(T_{wo} - T_{wi}\right)x}{A\left(T_2 - T_1\right)} \tag{3}$$

The thermal conductivity measurement depends on the measurement of the heat flux Q and temperature difference (T2 – T1). The axial flow method has been long established and has produced some of the most consistent and highest accuracy results. The present measurement focused mainly on reduction of radial heat losses in the axial heat flow developed through the specimen from the electrical heater mounted at one end.

RESULTS AND DISCUSSION

Microstructural Observations

Figure 3 shows typical micrographs for the microstruc- ture of the monolithic A356 alloy as well as the fabricated A356/Al_2O_3 MMNCs after heat treatment. It is clear from Figure 3(a) that the structure of the mono- lithic A356 Al alloy consists of primary α phase (white regions) and Al-Si eutectic structure (darker regions). Fine needle-like primary Si particulates were distributed along the boundaries of the α-Al dendrites. Figures 3(b) and 3(c) show micrographs of A356/Al_2O_3 MMNCs containing 3 vol% of Al2O3 nanoparticles having 60 and 200 nm, respectively. Clusters of Al_2O_3 nanoparticles in the microstructure of the A356/Al2O3 MMNCs were observed. Figure 3(d) shows high magnification optical micrograph of Al2O3/5 vol% (200 nm) MMNCs. It is clear that clusters of nanoparticles clusters are located inside the α grains as well as near the eutectic structure. Figure 4(a) shows high magnification SEM micrograph of the 5 vol% Al_2O_3 nanoparticles (200 nm) showing that nanoparticles are agglomerating near the Si particles of the eutectic structure. The XRD analysis for the nanoparticles is shown in Figure 4(b). Increasing the volume fraction of the Al_2O_3 nanoparticles dispersed inside the A356 alloy increases the agglomeration percent.

Figure 3. Optical micrographs for (a) A356 monolithic alu- minum alloy; (b) A356/3 vol% Al_2O_3 (60 nm) nanocompo- sites; (c) A356/3 vol% Al_2O_3 (200 nm) nanocomposites; (d) A356/5 vol% Al_2O_3 (200 nm); (e) A356/5 vol% Al_2O_3 (200 nm) nanocomposites.

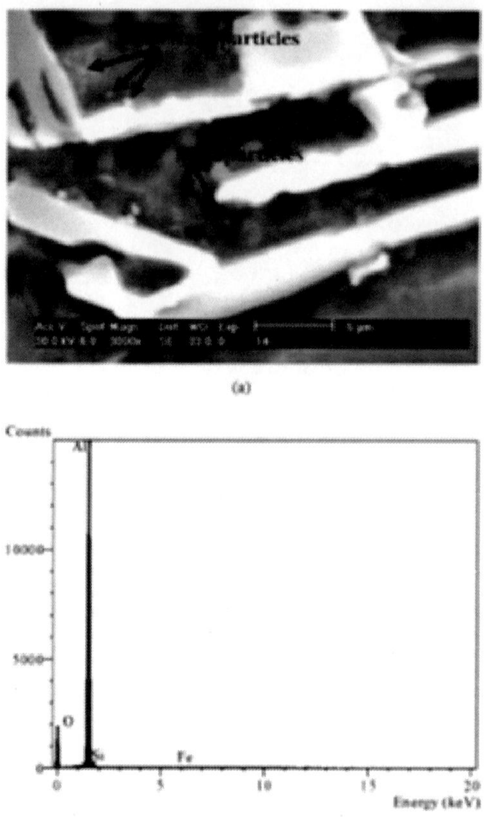

Figure 4. (a) High magnification SEM micrograph of the 5 vol% Al_2O_3 nanoparticulates (200 nm) showing that na- noparticulates are agglomerating near the Si particles of the eutectic structure; (b) XRD analysis for the particles shown in (a).

MMNCs containing 5 vol% of Al_2O_3 nanoparticles ex- hibited the highest agglomeration percent when com- pared with those containing 1 and 3 vol%.

Porosity measurements indicted that the MMNCs have porosity content lower than 2 vol%. Such low porosity content is attributed to the squeezing process that was carried out during the solidification of the MMNCs. In cast MMCs, there are several sources of gases. The occurrence of porosity can be attributed variously to the amount of hydrogen gas present in the melt, the oxide film on the surface of the melt that can be drawn into it at any stage of stirring and the gas being draw into the melt by certain stirring methods. Vigorously stirred melt or vortex tends to entrap gas and draw it into the melt. In- creasing the stirring time allows more gases to be entered into the melt and hence reduce the mechanical properties.

The amount of liquid inside the semi-solid slurry in- creases with increasing the temperature which on the other hand reduces the viscosity of the solid/ liquid slurry. Nanoparticle distribution in the A356 Al matrix alloy during the squeezing process depends greatly on the viscosity of the slurry and also on the characteristics of the reinforcement particles themselves, which influence the effectiveness of squeezing in to break up agglomerates and distribute particles. When the amount of liquid inside the slurry is large enough, the particles can be rolled or slid over each other and thus breaking up agglomerations and helping the redistribution of nanoparticles and improving the microstructure.

The Thermal Conductivity of MMNCs

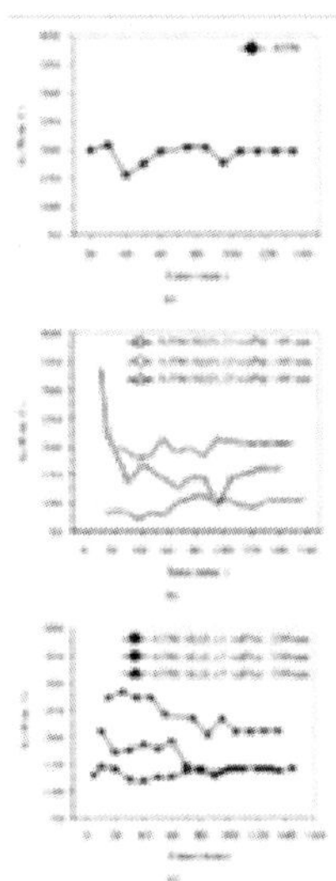

Figure 5. Variation of average thermal conductivity with the time for (a) the A356 monolithic alloys; (b) A356/Al_2O_3 60 nm) MMNCs and (c) A356/Al_2O_3 (200 nm) MMNCs.

Figure 5 shows the variation of the thermal conductivity against the time for different MMNCs specimens having different volume fractions and sizes of Al_2O_3 nanoparti- cles. The figure shows that there is a significant variation of the thermal conductivity during the first few minutes and it approaches to a relatively constant value after 120 minutes or two hours (Steady state condition). The A356/5 vol% contains Al_2O_3 nanoparticles of 200 nm and 60 nm exhibited the maximum and the minimum values of thermal conductivity of 213 and 105 W/m.K, respectively. The present experimental measurements show a signify- cant difference in the value of the average thermal con- ductivity for different A356/Al_2O_3 MMNCs. The steady state technique and the axial flow method are satisfactory methods for measuring the thermal conductivity of MMNCs.

The variation of the thermal conductivity of the A356/ Al_2O_3 MMNCs with the volume fraction of the nanoparticles at several nanoparticles sizes is shown in **Figure 6**. The thermal conductivities of the A356/Al_2O_3 MMNCs showed disturbing results. It is suggested that the clustering of the Al_2O_3 nanoparticles is the reason of such results.

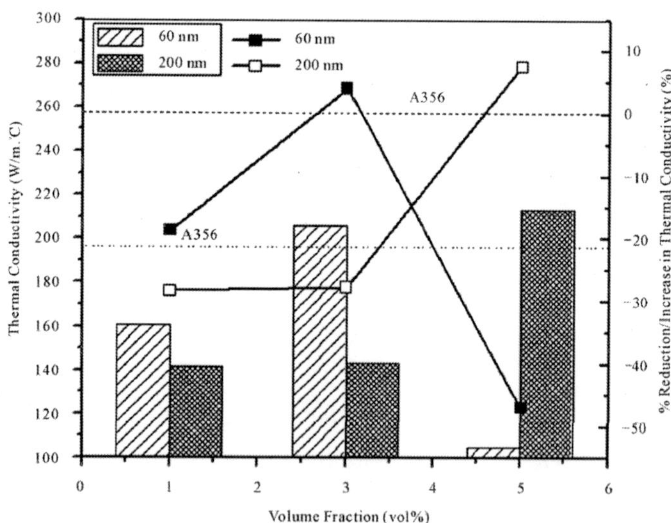

Figure 6. The average thermal conductivity of the A356/Al_2O_3 MMNCs.

Generally, the Al_2O_3 nanoparticles addition de- creased the thermal conductivity of the MMNCs when compared with the A356 monolithic alloy. The MMNCs containing up to 3% of 60 nm Al_2O_3 nanoparticles showed better thermal conductivities than those containing 200 nm nanoparticles. Only MMNCs containing 3%-60 nm and 5%-200 nm of Al_2O_3 nanoparticles showed practically the same thermal conductivity of A356. The reduction of the thermal conductivity of the MMNCs due to the addition of the Al_2O3 nanoparticles may attribute to the insulating nature of the Al_2O_3 nanoparticles itself and the increase of the porosity content with increasing the volume fraction of the nanoparticles. **Figure 6** shows also the percent of reduction/increase of the thermal conductivity of A356 due to the dispersion of Al_2O_3 nanoparticles having several sizes and volume fractions. The maximum percentage of reduction of thermal conductivity was about 47% for MMNCs containing 5 vol% of 60 nm particulates.

The reduction of the thermal conductivity due to the addition of ceramic nanoparticles was also reported by Ke Chu *et al.* [11]. They reported that the addition of carbon nanotubes (CNTs) showed no enhancement in overall thermal conductivity of the Cu composites due to the interface thermal resistance associated with the low phase contrast of CNT to copper and the random tube orientation. Besides, the composite containing 15 vol% CNTs led to a rather low thermal conductivity due possibly to the combined effect of unfavourable factors induced by the presence of CNT clusters, *i.e.* large porosity and lower effective conductivity of CNT clusters them- selves.

The thermal conductivity of the conventional aluminium based MMCs was investigated by many workers [12,13]. It has been found that the thermal conductivity of MMCs is mainly governed by the conductivity of the individual phases, their volume fraction and shape, and also by the size of the inclusion phase due to a finite metal/ceramic interface thermal resistance. Beside these factors, the results obtained from the current work showed that the agglomeration % of the particulates has a significant influence on the thermal conductivity of the MMCs. Increasing the volume fraction of the particulates increases the amount of agglomeration % of the particles which can lead to disturbing results of the average thermal conductivity of the MMCs. Particles free zones have higher thermal conductivity than the particles clustered zones. Another important factor that plays an important role on the thermal conductivity of the MMCs is the fabrication technique and its processing conditions. The casting techniques have several advantages than other fabrication techniques such as powder metallurgy. These advantages include lower cost and the capability of production of large components. However, MMCs cast components suffer from several defects such as particles segregation and high porosity content. The choice of the proper process parameters of the casting technique has a significant effect in reduction, but not elimination, of the particles segregation and porosity content. Squeezing the MMNCs during solidification helped in reduction of the porosity content of the MMNCs, however, it did not reduce the agglomeration % of the particles. The reduction of the agglomeration % of the particles can be achieved by the proper selection of the rheocasting process parameters which was the previous stage before the squeezeing process. The rheocasting process parameters include; the stirring speed; the shape, position and size of the stirrer; the stirring temperature; the stirring duration and pouring technique.

Electrical Conductivity of MMNCs

Figure 7 shows the variation of the electrical conductivity of the A356/Al_2O_3 MMNCs, in MS/m, with the volume fraction of the Al_2O_3 nanoparticles. The results revealed that the MMNCs showed lower electrical conductivities when compared with the A356 base alloy. Increasing the volume fraction reduces the electrical conductivity of the MMNCs. Moreover, the increasing the size of the nanoparticles from 60 to 200 nm reduced significantly the electrical conductivity of the MMNCs. The maximum reduction percent in the conductivity, with respect to the A356 monolithic alloy, was about 38% and exhibited by A356/Al_2O_3 MMNCs containing 5 vol% of nanoparticles having 200 nm.

The reduction of the electrical conductivity of the A356 due to the addition of the Al_2O_3 nanoparticles may be attributed to the clustering of the nanoparticles which are in fact insulations sites that reduce the electrical conductivity. Moreover, it has been observed that the nanoparticles are positioned and clustered at the grain boundaries of Al grains and decrease the electrical conductivity. The reduction of the electrical conductivity due to the addition of ceramic nanoparticles noticed in the current work has been observed also by Sheikh *et al.* [10]. In their work, carbon nanotubes were (CNT) added to pure copper and MMNCs were fabricated using powder metallurgy. The results revealed that increasing CNT contents in the Cu composites decreases the electrical conductivity. In contrast, *El-Mahallawi et al.* [14] reported that introducing Al_2O_3 nanoparticles to A356 alloy within the range 1 vol% tends to yield the alloy of highest electrical conductivity of the MMNCs. The morphology, and size of the eutectic silicon is believed to have an effect on the electrical conductivity also and any improvement on the eutectic morphology is believed to induce a significant effect on the electrical conductivity of the A356 alloy [14]. The significant effect induced on cast alloys by adding Al2O3 particulates has been dis- cussed and reported recently by some researchers. Generally, the electrical conductivity of the MMCs is mainly decided by the conductivity of the matrix alloy as well as the shape, size and volume fraction of hard ceramic alumina particles [15].

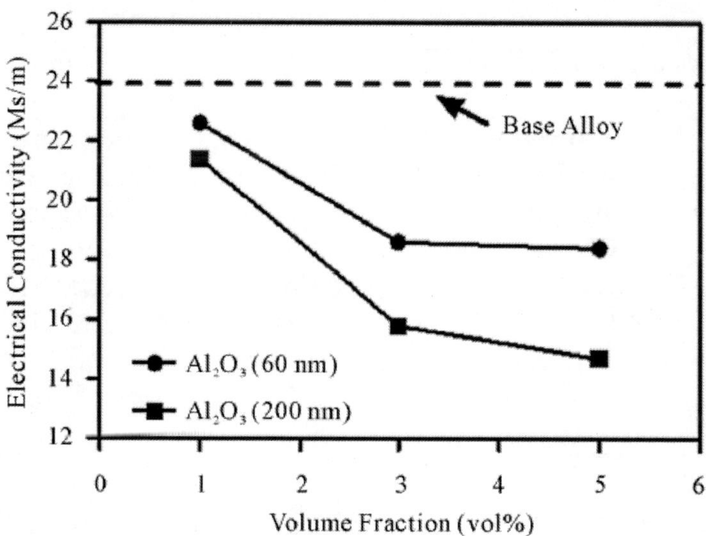

Figure 7. Variation of the electrical conductivity of the A356/Al2O3 MMNCs with the volume fraction of the Al2O3 nanoparticles.

CONCLUSIONS

Based on the results obtained from the current work, the following conclusions are drawn:

1) The A356 monolithic alloy exhibited better thermal conductivity than the MMNCs. The A356/Al2O3 MMNCs containing up to 3% of 60 nm Al2O3 nanoparticles showed better thermal conductivities than the MMNCs containing 200 nm nanoparticles.
2) The A356/Al2O3 MMNCs showed lower electrical conductivities when compared with the A356 base alloy. Increasing the volume fraction of the Al2O3 nanoparticles reduces the electrical conductivity of the MMNCs. Moreover, increasing the size of the Al2O3 nanoparticles from 60 to 200 nm reduced significantly the electrical conductivity of the MMNCs.
3) The maximum reduction percent in the thermal and electrical conductivities, according to the A356 monolithic alloy, were about 47% and 38%, respectively. These percentages were exhibited by A356/Al2O3 MMNCs containing 5 vol% of nanoparticles having 60 and 200 nm, respectively.

ACKNOWLEDGEMENTS

This work is supported by the King Abdel-Aziz City of Science and Technology (KACST) through the Science and Technology Center at King Khalid University (KKU), Fund (NAN 08-172-7). The authors thank both KACST and KKU for their financial support. Special Thanks to Prof. Dr. Saeed Saber, Vice President of KKU, Dr. Ahmed Taher, Dean of the Scientific Research at KKU, and Dr. Khaled Al-Zailaie, Dean of the faculty of engineering at KKU, for their support.

REFERENCES

1. ASM Handbook, "Composites," ASM International, Vol. 21, 2001.
2. N. Chawla and K. K. Chawla, "Metal Matrix Compos-ites," Springer Science and Business Media Inc., Berlin, 2006,
3. B. Cantor, F. P. E. Dunne and I. C. Stone, "Metal and Ceramic Matrix Composites," Taylor & Francis, New York, 2003.
4. J. Hashim, L. Looney and M. S. J. Hashmi, "Metal Matrix Composites: Production by the Stir Casting Method," Journal of Materials

Processing Technology, Vol. 92-93, 1999, pp. 1-7. doi:10.1016/S0924-0136(99)00118-1

5. K. Miwa and T. Ohashi, "Preparation of Fine SiC Particle Reinforced Al. Alloy Composites by Compocasting Process," Proceedings of the 5th Japan-U.S. Conference on Composite Materials, Tama City, Tokyo, 24-27 June 1990, pp. 355-362.
6. T. S. Mahmoud, F. H. Mahmoud, H. Zakaria and T. A. Khalifa, "Effect of Squeezing on Porosity and Wear Be-haviour of Partially Remelted A319/20vol%SiCpMMCs,"ProceedingsoftheInstitutionofMechanical Engineers, Part C: Journal of Mechanical Engineering Science, Vol. 222, No. C3, 2008, pp. 295-303. doi:10.1243/09544062JMES803
7. F. R. Rahmani and F. Akhlaghi, "Effect of Extrusion Temperature on the Microstructure and Porosity of A356-SiCp Composites," Journal of Materials Processing Technology, Vol. 187-188, 2007, pp. 433-436. doi:10.1016/j.jmatprotec.2006.11.077
8. A. Ansary Yar, M. Montazerian, H. Abdizadeh and H. R. Baharvandi, "Microstructure and Mechanical Properties of Aluminum Alloy Matrix Composite Reinforced with Nano-Particle MgO," Journal of Alloys and Compounds, Vol. 484, No. 1-2, 2009, pp. 400-404. doi:10.1016/j.jallcom.2009.04.117
9. J. Lan, Y. Yang and X. C. Li, "Microstructure and Mi-crohardness of SiC Nanoparticles Reinforced Magnesium Composites Fabricated by Ultrasonic Method," Materials Science and Engineering A, Vol. 386, No. 1-2, 2004, pp. 284-290.
10. S. M. Uddin, T. Mahmud, C. Wolf, C. Glanz, I. Kolaric, C. Volkmer, H. Höller, U. Wienecke, S. Roth and H.-J. Fecht, "Effect of Size and Shape of Metal Particles to Improve Hardness and Electrical Properties of Carbon Nanotube Reinforced Copper and Copper Alloy Composites," Composites Science and Technology, Vol. 70, No. 16, 2010, pp. 2253-2257.
11. K. Chu, Q. Y. Wu, C. C. Jia, X. B. Liang, J. H. Nie, W. H. Tian, G. S. Gai and H. Guo, "Fabrication and Effective Thermal Conductivity of Multi-Walled Carbon Nano-tubes Reinforced Cu Matrix Composites for Heat Sink Applications," Composites Science and Technology, Vol. 70, No. 2, 2010, pp. 298-304.
12. S. K. Chaudhury, A. K. Singh, C. S. Sivaramakrishnan and S. C. Panigrahi, "Effect of Processing Parameters on Physical Properties of Spray Formed and Stir Cast Al-2Mg-TiO2 Composites," Materials Science and Engi-neering A, Vol. 393, No. 1-2, 2005, pp. 196-203. doi:10.1016/j.msea.2004.10.010

13. M. Molina, J. Narciso, L. Weber, A. Mortensen and E. Louis, "Thermal Conductivity of Al–SiC Composites with Monomodal and Bimodal Particle Size Distribu-tion," Materials Science and Engineering A, Vol. 480, No. 1-2, 2008, pp. 483-488. doi:10.1016/j.msea.2007.07.026
14. I. S. El-Mahallawi, K. Eigenfield, F. Kouta, A. Hussein, T. S. Mahmoud, R. M. Ragaie, A. Y. Shash and W. Abou-Al-Hassan, "Synthesis and Characterization of New Cast A356/(Al2O3)P Metal Matrix Nano-Composites," ASME, Proceeding of the 2nd Multifunctional Nanocom-posites & Nanomaterials, International Conference & Exhibition, Organized by the American University in Cairo—AUC, in Collaboration with Cairo University, Sharm El Sheikh, Egypt, 11-13 January 2008.
15. L. Weber, J. Dorn and A. Mortensen, "On the Electrical Conductivity of Metal Matrix Composites Containing High Volume Fractions of Non-Conducting Inclusions," Acta Materialia, Vol. 51, No. 11, 2003, pp. 3199-3211. doi:10.1016/S1359-6454(03)00141-1

Chapter 9

CARBON NANOTUBES BASED 3-D MATRIX FOR ENABLING THREE-DIMENSIONAL NANO-MAGNETO-ELECTRONICS

Jeongmin Hong Mail, Eugenia Stefanescu, Ping Liang, Nikhil Joshi, Song Xue, Dmitri Litvinov,Sakhratkhizroev

Department of Electrical and Computer Engineering, Florida International University, Miami, Florida, United States of America

Department of Electrical Engineering, University of California Riverside, Riverside, California, United States of America

Seagate Technology, Bloomington, Minnesota, United States of America

MultiDimension, San Jose, California, United States of America

Department of Electrical Engineering, University of Houston, Houston, Texas, United States of America

Corresponding Author

Email: jehong@fiu.edu

ABSTRACT

This chapter describes the use of vertically aligned carbon nanotubes (CNT)-based arrays with estimated 2-nm thick cobalt (Co) nanoparticles deposited inside individual tubes to unravel the possibility of using the unique templates for ultra-high-density low-energy 3-D nano-magneto-electronic devices. The presence of oriented 2-nm thick Co layers within individual nanotubes in the CNT-based 3-D matrix is confirmed through VSM measurements as well as an energy-dispersive X-ray spectroscopy (EDS).

INTRODUCTION

The progress of planar silicon-based electronics technology – described by Moore's Law – is coming to an end. [1], [2] Gate-length scaling has pushed the gate-dielectric and junction technology to its physical limits and has resulted in a design in which billions of transistors are interconnected by tens of kilometers of wires packed into an area of square centimeters - new materials and processes have been introduced, but it is expected that the continued increase in interconnect wiring will ultimately lead to the demise of further CMOS scaling, quite apart from the very high power and associated thermal budgets that are required to drive such an architecture. [3] The electronics industry will be forced to meet the demand for additional functionality in a reduced footprint with revolutionary new electronic devices based on new state variable rather than electronic motion and by the extension of electronics into the third dimension. These next generation technologies will extend Moore's Law and create dynasties of low power electronics with greater capabilities at a fraction of the cost.

The leap in the integration of devices can be addressed by 3-D technology. [4] Conventional electronics is based on 2-D planar processes, but this is becoming prohibitively expensive as well as a barrier to performance. By stacking devices and interconnecting them in a 3-D arrangement, a huge leap in functional density is possible together with reduced power consumption. 3-D integration is a cornerstone of the coming revolution in electronics not only because of the possibility of high-density architectures but also because it will enable the introduction of disparate signals and new materials and devices. [5]–[11] The 3-D device architecture allows functionally integrating different materials in one system thus providing a platform to combine the advantages of nanomagnetic structures with the unprecedented electronic and thermal properties of new materials such as graphene and other Carbon based nano-formulations. Using a 3-D matrix of nanomagnetic nodes "connected" via magnetic fields promises also to substantially reduce the number of physical wires and interconnects (vias) which is another important roadblock in the current 2-D planar VLSI technology. Nanomagnetic devices can be partially polarized and therefore naturally allow for multi-valued signal coding. In nanomagnetic devices, multi-valued signals can be reliably achieved with high noise tolerance, no increased power consumption, and no significantly increased circuit complexity, thus making multi-valued logic a cost effective and power efficient signal processing solution, further increasing the computing capacity per unit area or volume.

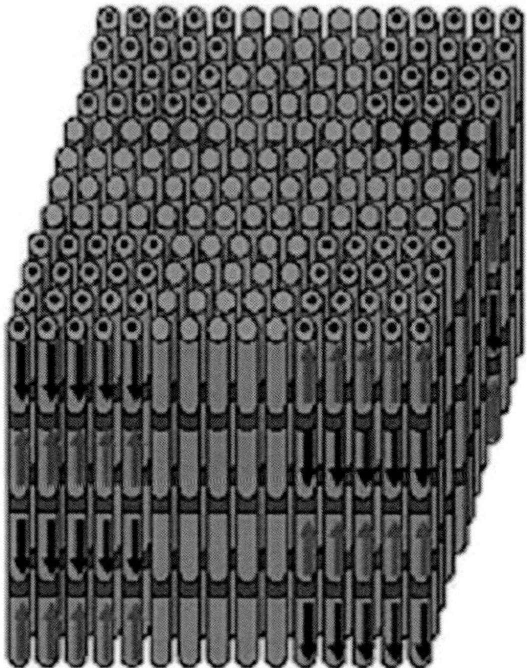

Figure 1. An array of vertically aligned CNTs with various magnetic materials deposited inside individual tubes. The magnetic anisotropy is shape-induced.

One of the innovative features of densely packed vertically aligned CNTs is the potential to be used as the 3-D matrices (templates) for enabling next-generation 3-D nano-magnetoelectronic devices. CNTs have been a promising candidate to build future electronic devices due to their many attractive properties. [12]–[19] In this paper, as a trivial example, a CNT matrix is used to control magnetic properties in nano scale.

RESULTS AND DISCUSSION

The inherent to the 3-D-matrix shape-induced magnetic anisotropy (due to the vertically aligned CNTs) ideally allows using any magnetic material in a specific orientation. The significance of this feature can be illustrated on the example of perpendicular magnetic recording (PMR) currently used as the core technology in magnetic disk drives. [20] The recording media used in PMR has an intrinsically induced anisotropy and consequently is severely material limited. Indeed, there are only a handful of materials used in the industry such as a hexagonal closed packed (hcc) phase of Co-based compounds, Co/Pt (or Pd) multilayer, high anisotropy L10 materials, and a few others. As a result, the optimization of magnetic properties of the recording media is challenging. On the contrary, with the use of CNT-based templates, ideally any magnetic material could be used with an effectively perpendicular orientation, as illustrated in Figure 1. Here, we show that even a magnetically "soft" material could be made relatively "hard" and oriented through inducing shape anisotropy. In this case, the saturation moment is defined by the material, while the anisotropy is due to the 3-D matrix architecture. More specifically, the shape-induced anisotropy can be controlled by controlling the size of the filled (by the magnetic materials) nanotubes, i.e. a longer nanotube would induce a stronger out-of-plane anisotropy.

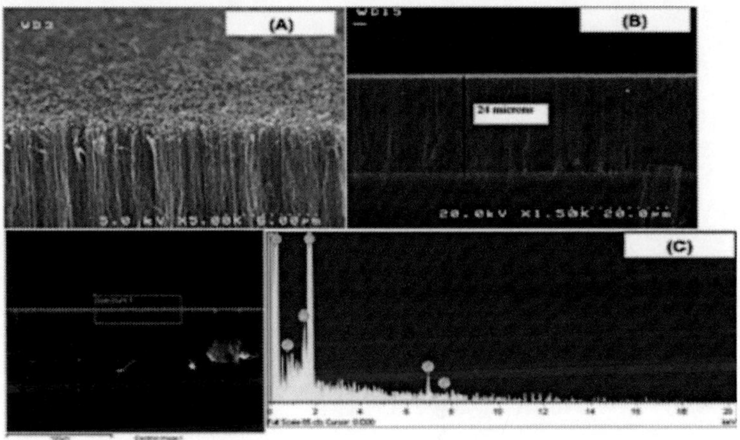

Figure 2. SEM images of (A) a vertically aligned CNT array and (B) a cross sectional image of an aligned CNT. (C) An energy-dispersive X-ray spectroscopy (EDS) of the composition of a vertically aligned CNT array. (The bottom left shows the targeted region for EDS pattern.).

There are many ways to deposit magnetic materials inside CNTs: 1) sputtering, 2) evaporation, 3) electroplating, and many others. In this particular experiment, we used porous Aluminum oxide (AAO) membranes to deposit a cobalt (Co) particles as a patterned array of catalyst sites. [21], [22] With such templates, with narrow distributions of diameters, Co nanoparticles were electrodeposited at the bottoms of openings to serve as catalysts in the subsequent pyrolysis step. Uniformly aligned carbon nanotubes were generated inside the pores of AAO through the catalytic pyrolysis of a hydrocarbon such as acetylene. The formed tubes had outer diameters equal to the inner diameters of the base pores. The desired cobalt particles were deposited within the nanotubes through the electroless deposition. Crystallite sizes in the range of 1.4 to 11.9 nm could be achieved. [23] Previous studies had shown that fcc, instead of hcp, was the stable and preferable structure of Co particles. [24], [25] It was found in some cases, Co would get inside the tubes and could occupy from 0 to 100 percent of a tube. Figure 2(A) and (B) represent SEM images of a vertically aligned high-aspect ratio CNT array. An energy-dispersive X-ray spectroscopy (EDS) of the composition of a vertically aligned CNT array is shown in Figure 2(C). The presence of Co is obvious from the spectrum.

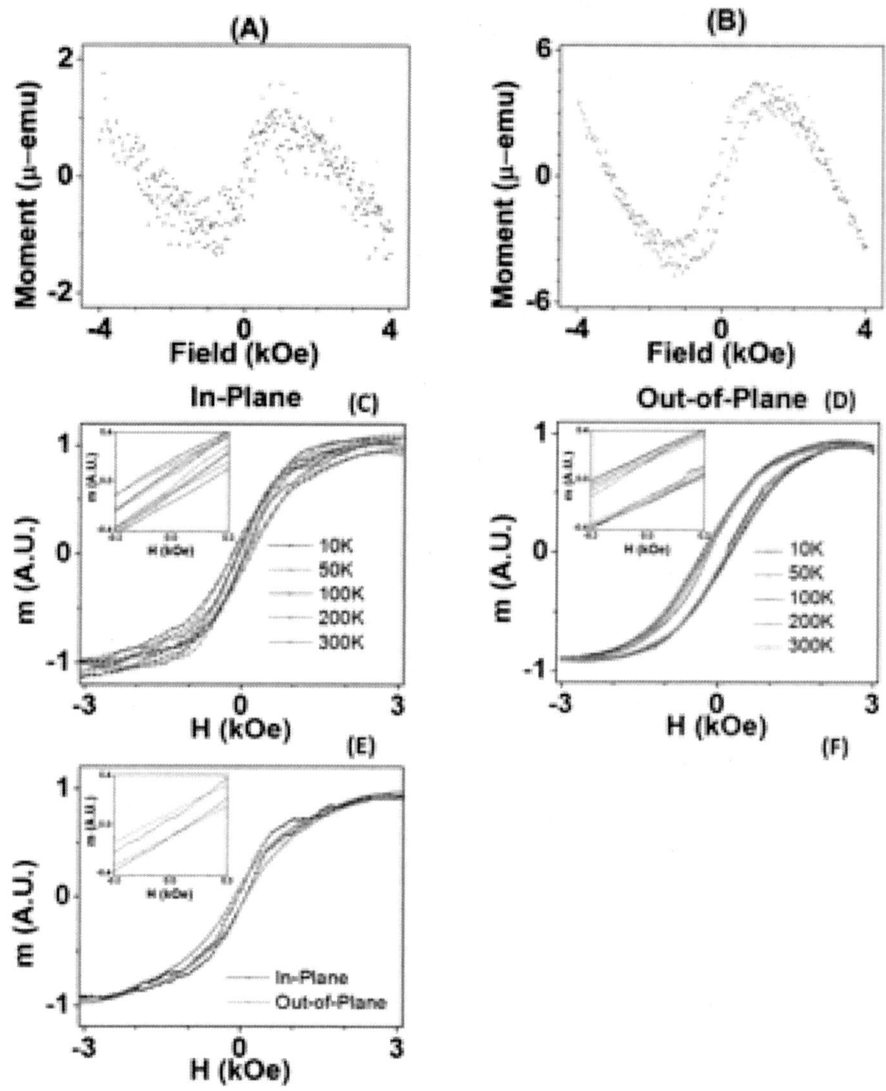

Figure 3. Magnetic moment (m) versus the applied field (H) hysteresis loops (A) along and (B) perpendicular to the plane directions at room temperature before subtracting diamagnetic background effects (originating from the substrates). (C) In-plane and (D) out-of-plane m-H loops for a range of temperature from 10 to 300 K after subtracting the diamagnetic background, (E) Room temperature m-H loops with out-of-plane and in-plane orientation, and (F) m-H loops from two directions before normalizing the magnetic moments. All the saturation values are above the value of the bulk fcc Co material [166 emu/g or 175 emu/g]. [27]–[29]

The magnetic moment of the CNT-based magnetic arrays was measured via vibrating sample magnetometry (VSM) PPMS (by Quantum Design, San Diego, CA). The measurements in two different directions, along and perpendicular to the plane are shown in Figures 3a and b, respectively. The more open hysteresis in the perpendicular direction and the anisotropy field higher than 1000 Oe (that is comparable to the saturation magnetization 4πMs of the order of 1000 emu/cc) indicate the orientation of the magnetization is tilted toward the perpendicular direction and is shape induced. For comparison, the magneto-crystalline anisotropy of Co doesn't exceed approximately 300 Oe, that is substantially smaller than the observed anisotropy value. Here, it should be mentioned that in the absence of CNT-tubes, the magnetization of Co would be in fcc (cubic) phase and preferentially in a plane direction.

Assuming that the anisotropy is indeed shape induced, i.e. $2\pi M_s$(assuming the cylindrical shape of a nanotube) one can estimate that the approximate volume of Co in the CNT sample is of the order 2×10^{-6} emu/10^3 emu/cc = 2×10^{-9}cc, which corresponds to the effective thickness of 2×10^{-9} cc/0.09 $cm^2 \sim 2\times10^{-7}$ cm = 2 nm.[26]

In summary, using an example of Co particles deposited inside individual CNTs, the study showed that densely packed vertically aligned CNTs could be indeed used a 3-D matrix for controlling the anisotropy direction of magnetic materials. The presence of oriented 2-nm thick Co layers within individual nanotubes in the CNT-based 3-D matrix was confirmed through VSM measurements as well as through an energy-dispersive X-ray spectroscopy (EDS).

MATERIALS AND METHODS

Deposition of Cobalt Nanoparticles

Porous Aluminum oxide (AAO) membranes were used to pre-deposit cobalt (Co) particles as a patterned array of catalyst sites. Uniformly aligned carbon nanotubes were generated inside the pores. The desired cobalt nanostructures were filled in the interior of the nanotubes using electroless deposition.

Vibrating Sample Magnetometry

M-H loop measurements were performed using the VSM option of a Quantum Design cryogenic physical property measurement system (PPMS) with a 9-Tesla superconducting magnet. Samples were mounted on a quartz paddle with regular disk holders, using GE-7031 varnish to withstand thermal cycling. To optimize the touchdown process, the samples were mounted with an upward offset of 35 mm.

SEM and EDS Measurements

The SEM and EDS study was performed with JEOL JSM 6330F.

ACKNOWLEDGMENTS

The authors would like to acknowledge the invaluable support from Motorola Corporation for many insightful discussions and help with the measurements.

AUTHOR CONTRIBUTIONS

Conceived and designed the experiments: JH ES PL SX DL SK. Performed the experiments: JH ES NJ. Analyzed the data: JH ES SK. Wrote the manuscript: JH SK.

REFERENCES

1. Wolf SA, Awschalom DD, Buhrman RA, Daughton JM, von Molnar S, et al. (2001) Spintronics: a spin-based electronics vision for the future. Science 294: 1488–95.
2. Tummala R (2006) Moore's law meets its match. IEEE Spectrum 43: 44–49.
3. Tanaka N, Yoshimira Y, Naito T, Miyazaki C, Nemoto Y, et al. (2005) Ultra-thin 3D-stacked SIP formed using room-temperature bonding between stacked chips. in Proceedings of the 55th Electronic Components and Technology Conference, Lake Buena Vista, FL, USA.
4. Emma PG, Kursun E (2008) Is 3D chip technology the next growth engine for performance improvement? IBM J. Res. & Dev. 52: 541–552.
5. Shen L, Chen W, Hung Y, Yang T, Leu F, et al. (2008) A scheme of array

memory stacking to the multi-channel solid state disk (SSD) applications: high speed, high reliability, and green compliance. in Proceedings of the 58th Electronic Components and Technology Conference, Lake Buena Vista, FL, USA.

6. Newman M, Muthukumar S, Schuelein M, Dambrauskas T, Dunaway P, et al. (2006) Fabrication and electrical characterization of 3D vertical interconnects. in Proceedings of the 56th Electronic Components and Technology Conference, San Diego, CA, USA.
7. Ranganathan N, Prasad K, Balasubramanian N, Pey KL (2008) A study of thermo-mechanical stress and its impact on through-silicon vias. J. Micromech. Microeng.18: 075018.
8. Lu KH, Ryu S, Zhao Q, Zhang X, Im J, et al. (2010) Thermal stress induced delamination of through silicon vias in 3-D interconnects. in Proceedings of the 60th Electronic Components and Technology Conference, Las Vegas, NV, USA.
9. Ramaswami S, Dukovic J, Eaton B, Pamarthy S, Bhatnagar A, et al. (2009) Process Integration Considerations for 300 mm TSV Manufacturing. IEEE Trans. Dev. Mater. Reliab. 9: 524–528.
10. Kikuchi H, Yamada Y, Ali A, Liang J, Fukushima T, et al. (2008) Tungsten through-silicon via technology for three-dimensional LSIs. Jpn. J. Appl. Phys. 47: 2801–2806.
11. Rimskog M (2007) Through wafer via technology for MEMS and 3D integration. in Proceedings of the 32nd IEEE/CPMT International Electronics Manufacturing Technology Symposium, San Jose, CA, USA.
12. Falvo MR, Clary GJ, Taylor RM, Chi V, Brooks FP, et al. (1997) Bending and buckling of carbon nanotubes under large strain. Nature 389: 582–584.
13. Jiang H, Liu B, Huang Y, Hwang KC (2004) Thermal expansion of single wall carbon nanotubes. J. Eng. Mater. Technol. 126: 265–270.
14. Wei BQ, Vajtai R, Ajayan PM (2001) Reliability and current carrying capacity of carbon nanotubes. Appl. Phys. Lett. 79: 1172–1174.
15. Collins PG, Hersam M, Arnold M, Martel R, AvourisPh (2001) Current saturation and electrical breakdown in multiwalled carbon nanotubes. Phys. Rev. Lett. 86: 3128–31.
16. Ragab T, Basaran C (2009) Joule heating in single-walled carbon nanotubes. J. Appl. Phys. 106: 063705.
17. Hone J, Whitney M, Piskoti C, Zettl A (1999) Thermal conductivity of single-walled carbon nanotubes. Phys. Rev. B 59: R2514.
18. Bom D, Andrews R, Jacques D, Anthony J, Chen B, et al. (2002) Thermogravimetric analysis of the oxidation of multi-walled carbon nanotubes: evidence for the role of defect sites in carbon nanotube chemistry. Nanolett. 2: 615–619.
19. Awano Y, Sato S, Nihei M, Sakai T, Ohno Y, et al. (2011) Carbon nanotubes for VLSI, interconnect and transistor applications. Proc. IEEE 98: 2015.
20. Iwasaki S, Nakamura Y (1977) An analysis for the magnetization mode for

high density magnetic recording. IEEE Trans. Magn. 13: 1272–1277.

21. Hong J, Denver H, Borca-Tasciuc D-A (2006) Fabrication and characterization of nickel nanowire polymer composites. MRS Proc. 963: 0963–Q20–60.
22. Li J, Moskovits M, Haslett T (1998) Nanoscaleelectroless metal deposition in aligned carbon nanotubes. Chem. Mater. 10: 1963–1967.
23. Matsuoka M, Imanishi S, Hayashi T (1998) Physical properties of electroless Ni-P alloy deposits from a pyrophosphate bath. Plat. and Surf. Finish. 76: 54–60.
24. McHenry ME, Majetich SA, Artman JO, DeGraef M, Staley SW (1994) Superparamagnetism in carbon-coated Co particles produced by the Kratschmer carbon arc process. Phys. Rev. B49: 11 358–11363.
25. Gong W, Li H, Zhao Z, Chen J (1991) Ultrafine particles of Fe, Co, and Ni: Ferromagnetic metals. J. Appl. Phys. 69: 5119–5121.
26. Abelmann L, Khizroev S, Litvinov D, Bain JA, Zhu J, et al. (2000) Micromagnetic simulation of an ultra-small single pole perpendicular write head. J. Appl. Phys., 87 9: 6636–8.
27. Nishikawa M, Kita E, Erata T, Tasaki A (1993) Enhanced magnetization in Co/MgO multilayer thin films. J. Magn. Magn.Mater. 126: 303–306.
28. Childress JR, Chien CL (1991) Reentrant magnetic behavior in fcc Co-Cu alloys. Phys. Rev. B 43: 8089–8093.
29. Chen JP, Sorensen CM, Klabunde KJ, Hadjipanayis GC (1995) Enhanced magnetization of nanoscale colloidal cobalt particles. Phys. Rev. B 51: 11527–11532.

Chapter 10

CARBON NANOTUBE SOLAR CELLS

Colin Klinger, Yogeshwari Patel, Henk W. Ch. Postma

Department of Physics and Astronomy, California State University Northridge, Northridge, California, United States of America

Corresponding Author

Email: postma@csun.edu

ABSTRACT

We present proof-of-concept all-carbon solar cells. They are made of a photoactive side of predominantly semiconducting nanotubes for photoconversion and a counter electrode made of a natural mixture of carbon nanotubes or graphite, connected by a liquid electrolyte through a redox reaction. The cells do not require rare source materials such as In or Pt, nor high-grade semiconductor processing equipment, do not rely on dye for photoconversion and therefore do not bleach, and are easy to fabricate using a spray-paint technique. We observe that cells with a lower concentration of carbon nanotubes on the active semiconducting electrode perform better than cells with a higher concentration of nanotubes. This effect is contrary to the expectation that a larger number of nanotubes would lead to more photoconversion and therefore more power generation. We attribute this to the presence of metallic nanotubes that provide a short for photo-excited electrons, bypassing the load. We demonstrate optimization strategies that improve cell efficiency by orders of magnitude. Once it is possible to make semiconducting-only carbon nanotube films, that may provide the greatest efficiency

INTRODUCTION

Solar cells have great potential as an alternative energy source because of the enormous amount of available energy and its distributed nature that may enable a distributed power generation grid[1]. However, for solar energy to be cost-effective on a utility scale, the price of purchase, installation, operation and maintenance over the lifetime of a solar panel per kWh generated must compare favorably to current power generation technology, which for fossil-fuel based generation is 0.03-0.05 $/kWh[2]. Improvements are being made to solar cells to 1) increase the efficiency, and 2) lower the price. For instance, solar concentrators are being developed that focus solar light reflecting off a large mirror on a solar cell with a smaller surface area. Multi-junction devices are being developed that use junctions between materials with different band gaps to capture a greater number of photons and limit loss of excess photon energy when the excited high-energy electron relaxes to the Fermi level.

Gratzel cells[3], also known as Dye-Sensitized Solar Cells (DSSCs), offer a particularly interesting path to cost-effective solar power. By sacrificing some efficiency but offering a greater reduction in cost, the total price per kWh can be reduced considerably. While this initial argument for DSSCs is very compelling, it is worth noting that the current state-of-the art DSSCs have efficiencies that rival their solid-state counterparts[4]–[6]. Another advantage of DSSCs is that they operate well in low-light and overcast conditions. DSSCs typically consist of a transparent semiconducting film on conducting glass that functions as a photo-active electrode (Figure 1a, top). A glass plate is coated with Pt and acts as the counter electrode (Figure 1a, bottom). Light-sensitive dye molecules are adsorbed on a semiconducting material on another slide and the assembly is immersed in an electrolyte, typically iodide-triiodide (I^-/I_3^-). An incoming photon with energy *hv* excites an electron from the dye into the conduction band of the semiconductor and it migrates to the bottom electrode. The electrolyte reduces the dye, creating triiodide ($3I^- \rightarrow I_3^- + 2e^-$). The electrons follow the external circuit through the load to the counter electrode. The triiodide migrates through the electrolyte to the Pt electrode and gets reduced, thereby completing the circuit. The transparent semiconductor is typically made of nanoporousTiO_2 . Using a nanoporous material significantly increases the surface area available for dye molecules but at the same time limits the electron migration rate. Different transparent semiconductors are being studied with higher mobility, such as nanowire-based electrodes[7],[8]. The liquid electrolyte is not very stable at the wide range of temperatures solar cells typically are exposed to, so high-mobility solids are being investigated as well[9],[10], culminating recently in a record 12% conversion efficiency[6]. Various dyes have been used in DSSCs, ranging from metal-free organic dyes[11]through highly efficient Ru-based organic dyes such as 'N3 dye'[12],[13] and 'black dye'[14]–[17]to engineered semiconductor quantum dots with a very high extinction coefficient[18]. C_{60} has been shown to work as a 'dye' as well[19],[20]. Carbon nanotubes (CNTs)[21],[22], offer a potentially cheaper and easier alternative to these materials. They are photo active, highly conductive, strong, and chemically inert. Carbon nanotubes can be synthesized in multiple ways such as chemical vapor deposition or laser ablation. The natural ratio of as-synthesized carbon nanotubes is 2/3 semiconducting to 1/3 metallic.

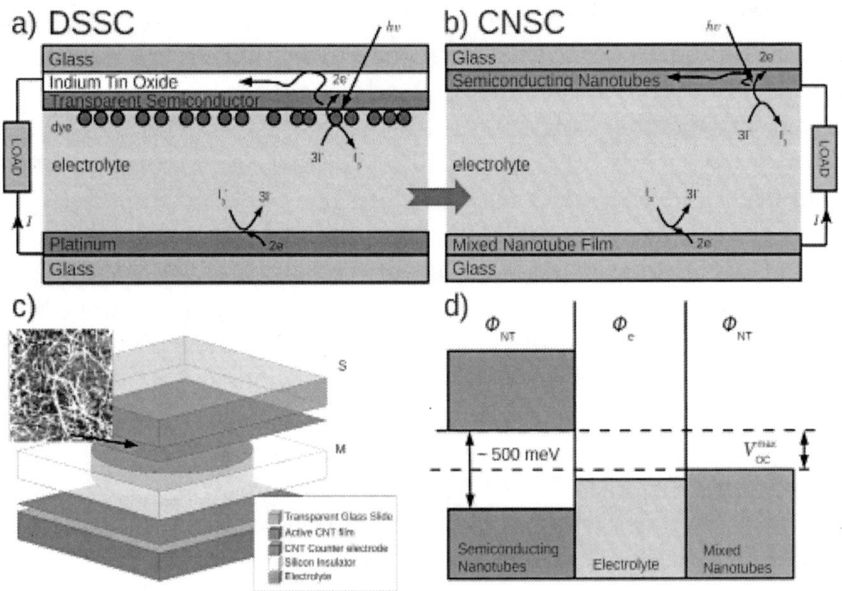

Figure 1. Carbon nanotube solar cells; comparison to Dye-Sensitized Solar Cells (DSSC), construction, and energeticts.

a) DSSC. b) Carbon Nanotube Solar Cell, CNSC. c) Layout of a CNSC. The top and bottom glass slides (light blue) are covered in carbon nanotube films which are electrically connected by the iodide-triiodide electrolyte (light red) that is contained by the silicone separator (white). The top film (green) is the photoactive electrode, while the bottom electrode (grey) is the counter electrode. The inset is an Atomic Force Micrograph of the height of a 2×2 m section of a carbon nanotube film. d) Band diagram of the CNSC.

Here, we present proof-of-concept solar cells that are entirely made of carbon nanotubes, carbon-nanotube-based solar cells (CNSCs,Figure 1b). They are a variation on the DSSC, and potentially offer many advantages beyond DSSCs. 1)No Dye**.** As these cells use semiconducting CNTs for photo conversion, they do not rely on dyes, which may bleach, severely limiting the useful life of DSSCs. 2)No Pt. Pt is often used as counter electrodes and their use in DSSCs represent an undesirable reliance on noble metals which may inhibit the use of DSSCs on a large, i.e. utility, scale. In addition, Pt has been reported to degrade due to the contact with the electrolyte[23].

Carbon nanotubes, in contrast, are chemically inert, and indeed show promising characteristics as counter electrodes[24]–[27]. 3)No In. As the carbon nanotube film itself is a transparent conductor, the use of a conducting coating made of, e.g. InSnO, is not required, eliminating the need for the exceedingly rare Indium. 4) The application of carbon nanotubes to the glass slides is a low temperature spray-coating process. In addition, these CNSCs multiply the advantages offered by DSSCs over single and multi-junction solar cells that require high-grade semiconductors and clean-room manufacturing. The use of low-grade materials and resulting projected significant reduction in cost of manufacturing potentially offsets the limited efficiency of these cells when relating the energy produced per dollar spent in manufacturing and installation.

In addition to CNT-only cells, we report on effiency improvement strategies, using different assembly techniques and using graphite (graphenium) counter electrodes. Graphite has no band gap, is extremely pliable, robust, and provides the ability to shrink the distance between it and the active semiconducting electrode. The cost, relative abundance, ease of introduction into the cell, and lack of need for spray deposition render graphite an attractive counter electrode material.

RESULTS

We present experimental demonstration of power generation obtained under ambient conditions at solar noon (seeMethodssection for details) of two types of cells.

1) *CNT-only cells*: cells are built with identical geometry but different CNT film compositions and thickness. This highlights how film composition affects cell performance (Figure 2a).
2) *Optimized cells*: cells are built with the same CNT film thickness and composition, but with differences in construction techniques to isolate its role in cell efficiency (Figure 2b–D).

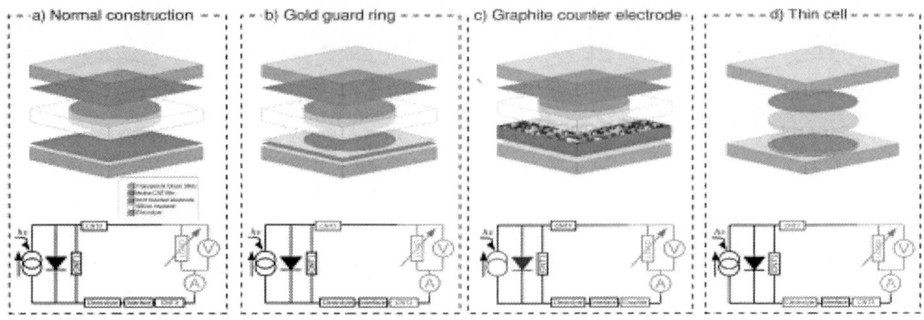

Figure 2. Layout of DSSCs and equivalent circuits.

Both basic DSSCs (a) and tested optimization strategies are (b–d) are depicted. The circuit diagram is modeled after[41]. The alternative construction techniques lead to changes in the cell's electrical model, which are highlighted in red.

CNT-Only Cells (Figure 1a)

The photocurrent *I* decreases linearly with increasing cell potential applied to the load*V*(Figure 3a). We extract the open-circuit voltage V_{OC} by extrapolating the I-V characteristic to I=0 and the short-circuit current I_{SC} by extrapolating to V=0 (Table 1). Both I_{SC} and V_{OC} of the*enriched*mixture cells increase with decreasing CNT coverage of the semiconducting active electrode. Similarly, the high-density cell of the*regular*mixture of nanotubes has a lower $I_{SC}$and V_{OC} than the low-density cell. The power transfer curves (Figure 3b) show a peak power transfer of P_{max} that occurs when the impedance of the load reachesR_{max} The low-density enriched as well as the low-density regular cells deliver more power to the load than their high-density counterparts. This is consistent with both $I_{SC}$and V_{OC} being larger.

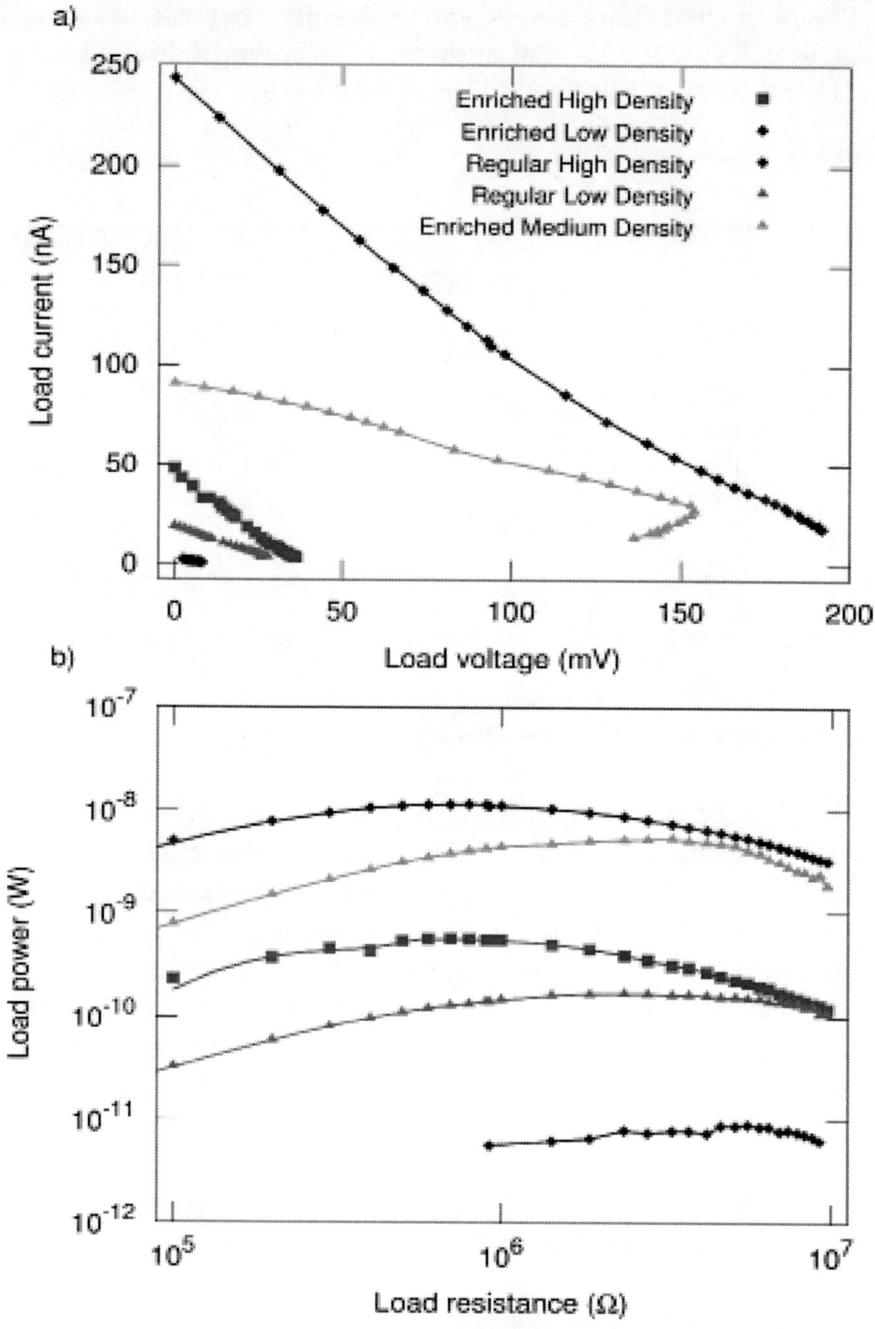

Figure 3. Electrical characteristics of the CNSCs.

The extracted parameters are presented inTable 2. a)$I-V$ characteristics of the cells as indicated. b) Power delivered to the load for all cells as described in the legend for a).

Table 1.Parameters of CNSCs.

	Cell Type	R–	Voc:	Isc	Pmax	Rmax	II_y	"111
		kW O	mV	nA	nW	MO	a.u.	a.u.
	Enriched High Density	3.4	43.6	47.0	0.57	0.71	521	58
(black)	Enriched Low Density	62.7	208.5	243.4	11.50	0.82	121	13
(blue)	Regular High Density	30.2	9.3	4.4	0.01	5.55	36	18
(magenta)	Regular Low Density	50.3	21.4	19.0	0.17	2.45	29	15
(green)	Enriched Medium Density	46.1	154.1	91.2	5.32	3.01	136	15

Optimized cells (Figure 2b–D)

We have studied cells with different construction techniques, using CNT electrodes from the same batch. Similar techniques were employed for data analysis as above (Figure 4,Table 2). The power transfer curves (Figure 4b) show a peak power transfer P_{max} at $R=R_{max}$.*Gold Guard Ring*. The presence of the gold guard ring increases I_{SC} by a factor ~0.25, whileV_{OC}remains approximately constant. R_{max} is lower by ~3.5 and P_{max} is ~2times greater. *Graphite Counter Electrode*. Both V_{OC} and I_{SC} are greater than the normally constructed cell and P_{max} is ~12times greater.*Thin Cell*. When the enriched side is facing the incident solar radiation ("up"), the power is slightly larger than when the regular side is facing the incident radiation ("down"). Both V_{OC} and I_{SC} are lower by factors of ~5 and ~3, which in itself is undesirable. However, optimum power transfer occurs at a much lower resistance.

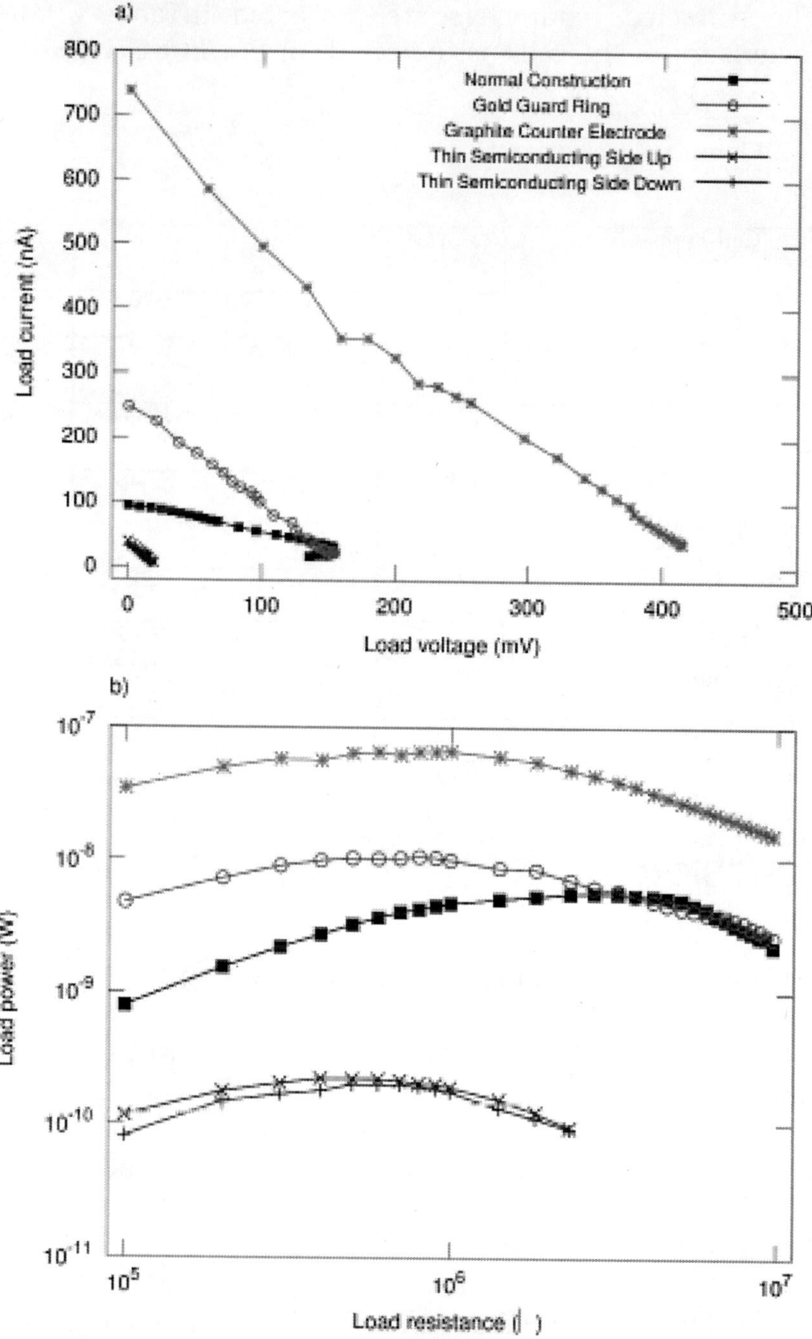

Figure 4. Optimization strategies for CNSCs.

The extracted parameters are presented inTable 1. a)$I - V$ characteristics of the cells as indicated. b) Power delivered to the load.

Table 2. Optimized CNSCs.

	Cell Type	VOC	'Sc	Pmax	Rmax
		mV	**nA**	**nW**	
•	Normal Construction	102.7	93.7	5.59	2.85
	Gold Guard Ring	129.5	249.5	10.46	0.81
	Graphite Counter	438.1	733.2	65.48	0.94
	Electrode				
x	Thin Semiconducting	22.0	38.3	0.22	0.47
	Side Up				
+	Thin Semiconducting	20.9	32.0	0.20	0.58
	Side Down				

DISCUSSION

Photocurrent Generation and Cell Voltage

The linear I-V characteristic phenomenologically indicates the source is purely resistive, and maximum power occurs when the load and source impedance are equal. A figure of merit for solar cells that describes how close its I-V characteristic is to the ideal shape is the fill factor FF which is defined as the ratio of P_{max} to the maximum power available with the corresponding ideal cell, $FF \equiv P_{max}/(V_{OC} I_{SC})$. It ranges from 0 to 1, where 1 indicates an ideal cell. Ideal cells can supply a constant voltage independent on the load resistance up to the maximum current, when the voltage drops quickly to 0. Deviations from the ideal fill factor of 1 are usually due to parasitic resistances, such as shunt and series resistances. Shunt resistances affect behavior in the (I-V) characteristic close to I_{SC}, while series resistances affect performance close to V_{OC}. For our cell, FF≈0.25. We argue that nanotube resistances 1–3 (Figure 2) are responsible for this. Effectively, it means that the diode in the circuit diagram can be neglected. To estimate the number of nanotubes, we use the measured sheet resistance presented inTable 1. Our CNT films are in the percolation limit[28],[29]. We can therefore use the scaling of sheet resistance with number of CNTs to extract the deposited volume of nanotube dispersion V, via:

$$R_{\square} \propto (V - V_C)^{-1.5},$$

Where V_C is the critical volume that determines the onset of conduction[28]. The volume can then be used to extract the surface density of metallic n_m and semiconducting nanotubes n_s. We assume that the metallic nanotubes dominate the sheet conductance, since their conductance G_m is much greater than semiconducting nanotubes G_s. This assumption holds provided the conductance ratio G_m / G_s of metallic to semiconducting nanotubes exceeds the semiconducting to metallic abundance ratio n_s/n_m . Single-molecule conductance studies of nanotubes indicate a conductance ratio of G_m / G_s ,[30],[31] which supports our assumption that metallic nanotubes dominate the sheet conductance. We anticipate that for more enriched semiconducting films than studied here, a more detailed analysis will be required that takes into account the nanotube-nanotube contact resistance as well[32],[33]. The current-generating capacity of our cells is proportional to the number of semiconducting nanotubes n_s. Combining both, the open-circuit condition corresponds to an ideal current source ($I \propto n_s$I) connected to CNT1 ($R_{CNT1} \propto R_{\square}$) and the voltage developed across it will be:

$$V_{OC} \propto n_s R_{\square}$$

and our data indeed approximately follows this scaling behavior (figure 5a). The outliers at low V_{OC} are CNT cells where both photoactive and counter electrode is coated with the same composition of carbon nanotubes. Both sides of the cell therefore create a photocurrent in opposite directions, but the light attenuation in the electrolyte breaks this symmetry and causes a directed current, albeit a smaller one and with a smaller voltage. The enriched cells further tilts the balance in favor of the photoactive side, leading to a V_{OC} that is closer to that expected from the amount of nanotube material deposited on the active side alone.

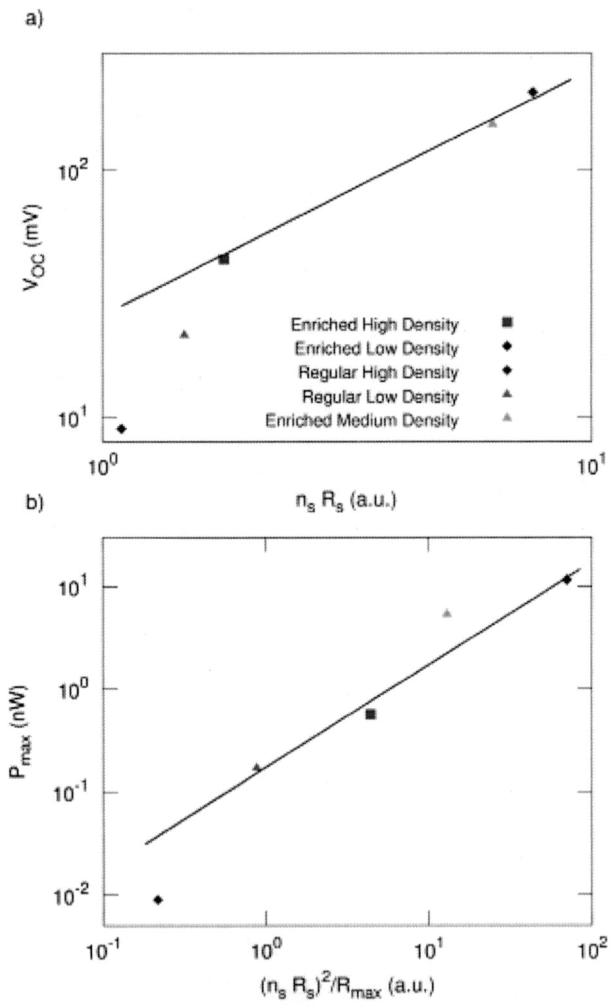

Figure 5. Scaling analysis of CNSC performance

The CNSCs characteristics are determined by the metallic and semiconducting carbon nanotube densities, with symbols corresponding to the cells as in the legend for Figure 2. a) $V_{OC} \propto n_s R_{\square}$. b) $P_{max} \propto (n_s R_{\square})^2 / R_{max}$.

Our cells have rather large output impedances and cannot maintain constant voltage over a larger range of load impedance. The cell output resistance can be reduced considerably by changing the aspect ratio of the cell, or connecting many cells in parallel. The output voltage can be held constant by a voltage-regulation circuit. However, there are many applications that do not require a low output impedance and would therefore work well with CNSCs, e.g. driving an LCD display or an E-Ink screen.

Maximum Open Circuit Voltage

The band gap of semiconducting carbon nanotubes is related to the nanotube's diameter d through:

$$E_g = 2\gamma_0 \frac{a_{CC}}{d},$$

Where $\gamma_0 = 2.45$eV is the nearest-neighbour overlap integral and $a_{CC}$is the carbon-carbon distance [30], [34]–[36]. Note that the band gap of semiconducting nanotubes does not depend on the chiral angle. The band gap for a 1.5 nm nanotube, the average diameter of our nanotube material, therefore amounts to~ 500meV. The band diagram is drawn inFigure 1d. The 'work function' of the electrolyte is$\Phi_e = 4.85$eV[37], while the work function for carbon nanotubes is$\Phi_{NT} = 4.5$eV[36]. We therefore expect the maximum attainable open circuit voltage$V_{OC}^{max} \sim 200 - 290\ mV$, where the range indicates variations due to the diameter. We observe a maximum voltage of ~ 200mV. This is expected as nanotubes with a slightly smaller band gap will 'short out' the effect of nanotubes with a slightly larger bandgap.

Solar Power Generation

The maximum power delivered to the load is a function of V_{OC} as well as the other resistances in the cell. The composite resistance of the cell is measured by determining at what value R_{max} of R_L maximum power transfer occurs. Since we can model our cells as a voltage source with source voltage $V_{OC} \propto n_s R_{\square}$, the maximum power available is expected to be $P_{max} \propto V_{OC}^2 / R_{max}$, or

$$P_{max} \propto \frac{(n_s R_{\square})^2}{R_{max}},$$

and indeed the maximum power appears to follow this behavior approximately (Figure 5b).

In summary, both P_{max} and V_{OC} behave according to our model that describes the role of film's resistive properties on cell performance.

Optimized Cell Designs

The *gold guard ring* causes 1) an increase in I_{SC}, 2) an increase in P_{max}, 3) a decrease in R_{max}, and 4) hardly any change in V_{OC}. We believe this is due to a reduction of the resistance of the nanotube film in contact with the silicone insulator (CNT2 and CNT3, Figure 2). The pressure of the insulator on the nanotube film as well as residual shear force during assembly may cause a perturbation of the nanotube film. In addition, the gold lowers the resistance of that part of the nanotube film. Both effects combined act to lower $R_{CNT2} + R_{CNT3}$. As the open-circuit voltage is independent of $R_{CNT2} + R_{CNT3}$, the open-circuit voltage should be unaffected by this improvement in design, and indeed we observe V_{OC}(Normal)~ V_{OC}(Gold Guard Ring). The reduction of R_{max} is also explained by the reduction of $R_{CNT2} + R_{CNT3}$, and that, in turn, explains the increase of both I_{SC} and P_{max}.

The employment of a *graphite counter electrode* instead of a carbon-nanotube counter electrode not only improves I_{SC}, but also V_{OC}. The increase in V_{OC} is due to the use of graphite instead of carbon nanotubes. The increase in I_{SC} is due to the lower sheet resistance of the graphite as compared to a carbon-nanotube film. The magnitude of the improvement is similar to that accomplished with the gold guard ring improvement, as the maximum power transfer occurs at approximately the same load resistance [R_{max}(Gold Guard Ring)~ R_{max}(Graphite Counter Electrode)]. Graphite is preferred over gold, naturally, to reduce cell cost. Both effects combine to increase the power output of the cell by a factor of ~ 12.

The reduction of the distance between active and counter electrode for the thin cells can reasonably be expected to lower the resistance of the electrolyte. In addition, as these cells were constructed without a silicone separator, we observe that the maximum power transfer occurs at a much lower resistance. This is due to the absence of the disruptive effect of the silicone separator, which role was elucidated by the study of the gold guard ring device above. The reduction of the electrolyte chamber thickness also has an adverse effect. The electrolyte absorbs less solar radiation than with thicker devices. Therefore, both the enriched (photo active) side, as well as the regular mixture side creates a photo current. However, both sources have opposite polarities, causing the effective open-circuit voltage to be reduced as we indeed observe.

Efficiency

The average solar flux during testing was 770 W/m2, and the greatest solar power generation was attained with the graphite counter electrode and enriched medium-density CNT active electrode. The efficiency of that cell was 1.8×10^{-5}. Compared to the all-CNT construction, an improvement of more than a factor 10 was attained. If a cell were constructed with the graphite counter electrode and the low-concentration CNT enriched active electrode, an increase of power by a factor 2 is anticipated. This can be deduced by comparison of the medium density enriched cell to the low density enriched cells with the regular construction. As the graphite counter electrode lowered the output resistance by a factor ~ 3, the power output may be larger by a factor 3 as well. Further improvements may be obtained by changing the aspect ratio of the solar cell. In the design reported here, we used effectively square films. Changing the cell design by making the cells wider, will lower the resistance further. An aspect ratio of 10 can then reduce the film resistance by a factor of 10, causing a reduction of R_{max}, which will improve P_{max}. Our thin cell results indicate that the largest resistance is due to the nanotube film, we therefore believe the efficiency increase with this improvement may be as large as 10-fold. We believe the greatest efficiency increase may be obtained by using CNT source material with a greater fraction of semiconducting nanotubes. The films we used had 90% semiconducting nanotubes and 10% metallic nanotubes. As we argued above, the semiconducting nanotubes provide the photo-generated current, but the metallic nanotubes short the load. If one were to use 99% semiconducting films, the amount of nanotubes could be increased by a factor 10, while still maintaining the same number of metallic nanotubes. As metallic nanotubes are more conductive than semiconducting nanotubes, we assume that the number of semiconducting nanotubes can be increased by this factor 10 without affecting $R_{\square}$ and R_{max}. Future generation cells can then reasonably be expected to deliver 100 times more power, due to the increase of n_s by a factor 10 (equation 4). However, at a certain abundance factor of semiconducting to metallic nanotubes, this argument will not hold any longer. Combining all of these improvements may lead to an efficiency of $0.8-5\%$, where the lower bound is a conservative estimate that every improvement will only contribute half we argued above. We hope the studies reported here will motivate further development of methods to create highly-enriched semiconducting CNT source material cost effectively at a large scale.

MATERIALS AND METHODS

Cell Construction

The *enriched* CNSCs (Figure 1) consist of a transparent glass slide covered with an enriched mixture of 90% semiconducting and 10% metallic nanotubes (IsoNanotubes-S 90% Powder, Nano Integris Inc.). These nanotubes have a diameter of $1.2-1.7$ nm and a length of $0.1-4\mu$m. Below this is a silicone insulator with a hole filled with electrolyte (iodide-triiodide, Solaronix). The electrolyte is in contact with both carbon nanotube films and acts to reduce the photo-active side as well as close the electrical circuit at the counter electrode. At the bottom is a glass slide covered with a regular mixture of $2/3$semiconducting and $1/3$metallic nanotubes that acts as a counter electrode (Unidym, lot PO-325, formerly Carbon Nanotechnologies Inc.). These nanotubes have a diameter of$0.8-1.2$nm and a length of$100-1000$ m. Dispersion of carbon nanotubes were made by ultrasonic agitation in 1, 2-dichloroethane for 1 h for the regular mixture and 4 h for the enriched mixture. The dispersion was spray painted with an air brush onto glass substrates in a vented cylindrical enclosure. The slides were rotated while spraying to obtain uniform coverage. Subsequently, the glass slides were heated on a hot plate to evaporate any residual solvent. The resulting film is similar to the well-known bucky paper and it has metallic properties[38]–[40]. The final solar cell has an exposed surface area of~ 4.8mm2with a distance of~ 2.5mm between the electrodes. The glass slides are 1 mm thick and did not have a conducting coating prior to carbon nanotube application.

Gold Guard Ring (Figure 2b)

A mask the size of the opening containing the electrolyte was placed onto the glass slide after CNT deposition, followed by Au deposition. This procedure prevents degration of the metal due to the electrolyte contact.

Graphite Cell (Figure 2c)

A cell using graphite (graphenium) as the counter electrode was created. The graphite cell counter electrode construction consists of the same steps for deposition of semiconducting CNTs. A PDMS (Slygard 184 Silicone Elastomer, Dow Corning Corp.) plastic mold with a circular depression was created to house pieces of graphite of different heights. A wire is placed through the PDMS at the height of the bottom of the depression. After graphite deposition the PDMS was filled with electrolyte and the active semiconducting electrode was placed on top of the cell.

Thin Cell (Figure 2d)

A cell with a separation of about~ 0.65mm between the active 90% semiconducting electrode and a regular CNT counter electrode was created. A piece of 1 mm thick glass was locally machined to create a central depression with a connection to a ramped section. The glass was cleaned and masked in the non-machines areas. The glass was sprayed with the $2/3$ semiconducting and $1/3$ metallic nanotubes mixture. A second piece of glass was masked with the same pattern as the machined glass piece and sprayed with 90% semiconducting CNTs. The two nanotube electrodes were connected to external electrodes and the cell was filled with electrolyte and sealed with liquid silicone sealant. The liquid silicone was allowed to dry and harden creating a seal. The thin cell has an exposed surface area of $\sim 48.9 mm^2$ with a distance of $0.55-0.75$ mm between the electrodes.

CNT film preparation and characterization

The spray-painted CNT slides were imaged with an Atomic Force Microscope (Dual-Scan AFM, Pacific Nanotechnology, USA) to determine the coverage (Figure 1). In addition, a probe station was used to measure the sheet resistance $R_{\square}(\Omega/\square)$ of the CNT films, by analyzing the distance dependence of the two-terminal resistance R as a function of probe separation L on a semilog scale and performing a least-squares fit to

$$R = \frac{R_{\square}}{\pi}\log\left(\frac{L}{d}\right),$$

Where d is the probe tip diameter. The counter electrodes used in this study were all obtained from the same batch in order to ensure uniformity and their sheet resistance was $R_{\square} \approx 50 k\Omega/\square$.

Solar Power Generation measurements

The assembled cells were connected to a load resistor R_L that was varied from 0 to 10 MΩ through a current amplifier and the voltage V across and current I through it are measured as a function of R_L (Figure 2). The cells were pointed straight at the sun and were measured in Northridge, CA (visibility: 10 miles, Latitude$=34°$N) at solar noon from Dec 2010 through April 2011. The sun's altitude β was between32.5°and 65.7°, yielding an air mass of $AM = 1/\sin\beta \approx 1.7$. The average solar flux was 770 W/m2. To minimize the effect of variability in solar conditions and cell assembly, devices were fabricated with large variations in carbon nanotube concentrations to highlight its effect on cell performance.

CNTs were created in batch operations, providing the ability to test various parameters and the resistances of electrodes used in experiments.

ACKNOWLEDGMENTS

We thank Konstantin Daskalov, Hiral Patel, AshkanForoughi, Daniel Frochtzwajg, KaroKarapetyan, and Michael Darling for experimental assistance. This work is dedicated to the memories of Michael Dickson and Andrew Wieting.

AUTHOR CONTRIBUTIONS

Conceived and designed the experiments: CK YP HP. Performed the experiments: CK YP HP. Analyzed the data: CK YP HP. Contributed reagents/materials/analysis tools: CK YP HP. Wrote the paper: CK YP HP.

REFERENCES

1. Lewis NS (2007) Toward Cost-Effective solar energy use. Science 315: 798–801.
2. Goldemberg J, Johansson TB, Anderson D (2004) World energy assessment: overview 2004 update. United Nations.
3. Gratzel M (2004) Conversion of sunlight to electric power by nanocrystalline dye-sensitized solar cells. Journal of Photochemistry and Photobiology A 164: 3–14.
4. Gao F, Wang Y, Zhang J, Shi D, Wang M, et al. (2008) A new heteroleptic ruthenium sensitizer enhances the absorptivity of mesoporoustitania _lm for a high efficiency dye-sensitized solar cell. Chemical Communications 2635–2637.
5. Shi D, Pootrakulchote N, Li R, Guo J, Wang Y, et al. (2008) New efficiency records for stable Dye-Sensitized solar cells with Low-Volatility and ionic liquid electrolytes. The Journal of Physical Chemistry C 112: 17046–17050.

6. Yella A, Lee H, Tsao HN, Yi C, Chandiran AK, et al. (2011) Porphyrin-Sensitized solar cells with cobalt (II/III)Based redox electrolyte exceed 12 percent efficiency. Science 334: 629–634.
7. Baxter JB, Aydil ES (2005) Nanowire-based dye-sensitized solar cells. Applied Physics Letters 86: 053114.
8. Law M, Greene LE, Johnson JC, Saykally R, Yang P (2005) Nanowire dye-sensitized solar cells. Nature Materials 4: 455–459.
9. Bach U, Lupo D, Comte P, Moser JE, Weissortel F, et al. (1998) Solid-state dye-sensitized mesoporousTiO 2 solar cells with high photon-to-electron conversion efficiencies. Nature 395: 583–585.
10. Saito Y, Kitamura T, Wada Y, Yanagida S (2002) Poly (3, 4-ethylenedioxythiophene) as a hole conductor in solid state dye sensitized solar cells. Synthetic Metals 131: 185–187.
11. Horiuchi T, Miura H, Sumioka K, Uchida S (2004) High efficiency of Dye-Sensitized solar cells based on Metal-Free indoline dyes. Journal of the American Chemical Society 126: 12218–12219.
12. Nazeeruddin MK, Kay A, Rodicio I, Humphry-Baker R, Mueller E, et al. (1993) Conversion of light to electricity by cis-X2bis(2,2′-bipyridyl-4,4′-dicarboxylate)ruthenium(II)charge-transfer sensitizers (X = cl-, br-, i-, CN-, and SCN-) on nanocrystalline titanium dioxide electrodes. Journal of the American Chemical Society 115: 6382–6390.
13. Kohle O, Ruile S, Gratzel M (1996) Ruthenium(II) Charge-Transfer sensitizers containing 4,4-Dicarboxy-2,2-bipyridine. synthesis, properties, and bonding mode of coordinated thio- and selenocyanates. Inorganic Chemistry 35: 4779–4787.
14. Nazeeruddin MK, Pchy P, Graetzel M (1997) Efficient panchromatic sensitization of nanocrystallineTiO 2 films by a black dye based on a trithiocyanatoruthenium complex. Chemical Communications 1997: 1705–1706.
15. Aiga F, Tada T (2003) Molecular and electronic structures of black dye; an efficient sensitizing dye for nanocrystalline TiO2 solar cells. Journal of Molecular Structure 658: 25–32.
16. Boschloo G, Lindstrom H, Magnusson E, Holmberg A, Hagfeldt A (2002) Optimization of dye-sensitized solar cells prepared by compression method. Journal of Photochemistry & Photobiology, A: Chemistry 148: 11–15.
17. Wang ZS, Yamaguchi T, Sugihara H, Arakawa H (2005) Significant efficiency improvement of the black dye-sensitized solar cell through protonation of TiO2 films. Langmuir 21: 4272–4276.
18. Vogel R, Hoyer P, Weller H (1994) Quantum-Sized PbS, CdS, Ag2S, Sb2S3, and Bi2S3 particles as sensitizers for various nanoporous Wide-Bandgap semiconductors. The Journal of Physical Chemistry 98: 3183–3188.
19. Li C, Mitra S (2007) Processing of fullerene-single wall carbon nanotube complex for bulk hetero-junction photovoltaic cells. Applied Physics Letters 91: 253112.

20. Li C, Chen Y, Wang Y, Iqbal Z, Chhowalla M, et al. (2007) A fullerenesingle wall carbon nanotube complex for polymer bulk heterojunction photovoltaic cells. Journal of Materials Chemistry 17: 24062411.
21. Iijima S, Ichihashi T (1993) Single-shell carbon nanotubes of 1-nm diameter. Nature 363: 603–605.
22. Bethune DS, Klang CH, de Vries MS, Gorman G, Savoy R, et al. (1993) Cobalt-catalysed growth of carbon nanotubes with single-atomic-layer walls. Nature 363: 605–607.
23. Koo B, Lee D, Kim H, Lee W, Song J, et al. (2006) Seasoning effect of dye-sensitized solar cells with different counter electrodes. Journal of Electroceramics 17: 79–82.
24. Trancik JE, Barton SC, Hone J (2008) Transparent and catalytic carbon nanotube films. Nano Letters 8: 982–987.
25. Ramasamy E, Lee WJ, Lee DY, Song JS (2008) Spray coated multi-wall carbon nanotube counter electrode for tri-iodide reduction in dye-sensitized solar cells. Electrochemistry Communications 10: 1087–1089.
26. Hwang S, Moon J, Lee S, Kim D, Lee D, et al. (2007) Carbon nanotubes as counter electrode for dye-sensitised solar cells. Electronics Letters 43: 1455–1456.
27. Kang M, Han Y, Choi H, Jeon M (2010) Two-step heat treatment of carbon nanotube based paste as counter electrode of dye-sensitised solar cells. Electronics Letters 46: 1509–1510.
28. Hu L, Hecht DS, Gruner G (2004) Percolation in transparent and conducting carbon nanotube networks. Nano Letters 4: 2513–2517.
29. Zhou Y, Hu L, Gruner G (2006) A method of printing carbon nanotube thin films. Applied Physics Letters 88: 123109.
30. Tans S, Verschueren A, Dekker C (1998) Room-temperature transistor based on a single carbon nanotube. Nature 393: 49–52.
31. Postma HWC, Teepen T, Yao Z, Grifoni M, Dekker C (2001) Carbon nanotube Single-Electron transistors at room temperature. Science 293: 76–79.
32. Yao Z, Postma HWC, Balents L, Dekker C (1999) Carbon nanotube intramolecular junctions. Nature 402: 273–276.
33. Postma HWC, de Jonge M, Yao Z, Dekker C (2000) Electrical transport through carbon nanotube junctions created by mechanical manipulation. Physical Review B 62: R10653–R10656.
34. Martel R, Schmidt T, Shea HR, Hertel T, Avouris P (1998) Single- and multi-wall carbon nanotube field-effect transistors. Applied Physics Letters 73: 2447.
35. Odom TW, Huang J, Kim P, Lieber CM (1998) Atomic structure and electronic properties of single-walled carbon nanotubes. Nature 391: 62–64.
36. Wildoer JWG, Venema LC, Rinzler AG, Smalley RE, Dekker C (1998) Electronic structure of atomically resolved carbon nanotubes. Nature 391: 59–62.

37. Tang Y, Lee C, Xu J, Liu Z, Chen Z, et al. (2010) Incorporation of graphenes in nanostructured TiO2 films via molecular grafting for Dye-Sensitized solar cell application. ACS Nano 4: 3482–3488.
38. Mickelson ET, Huffman CB, Rinzler AG, Smalley RE, Hauge RH, et al. (1998) Fluorination of single-wall carbon nanotubes. Chemical physics letters 296: 188–194.
39. Rinzler AG, Liu J, Dai H, Nikolaev P, Huffman CB, et al. (1998) Large-scale purification of single- wall carbon nanotubes: process, product, and characterization. ApplPhysA 67: 29–37.
40. Vigolo B, Penicaud A, Coulon C, Sauder C, Pailler R, et al. (2000) Macroscopic fibers and ribbons of oriented carbon nanotubes. Science 290: 1331–1334.
41. Han L, Koide N, Chiba Y, Islam A, Mitate T (2006) Modeling of an equivalent circuit for dye-sensitized solar cells: improvement of efficiency of dye-sensitized solar cells by reducing internal resistance. ComptesRendusChimie 9: 645–651.

Chapter 11

DROSOPHILA EMBRYOS AS MODEL TO ASSESS CELLULAR AND DEVELOPMENTAL TOXICITY OF MULTI-WALLED CARBON NANOTUBES (MWCNT) IN LIVING ORGANISMS

Boyin Liu, Eva M. Campo, Torsten Bossing

School of Biological Sciences, University of Bangor, Bangor, United Kingdom
School of Electronic Engineering, University of Bangor, Bangor, United Kingdom
Corresponding Author
Email: torsten.bossing@plymouth.ac.uk

ABSTRACT

Different toxicity tests for carbon nanotubes (CNT) have been developed to assess their impact on human health and on aquatic and terrestrial animal and plant life. We present a new model, the fruit fly Drosophila embryo offering the opportunity for rapid, inexpensive and detailed analysis of CNTs toxicity during embryonic development. We show that injected DiI labelled multi-walled carbon nanotubes (MWCNTs) become incorporated into cells in early Drosophila embryos, allowing the study of the consequences of cellular uptake of CNTs on cell communication, tissue and organ formation in living embryos. Fluorescently labelled subcellular structures showed that MWCNTs remained cytoplasmic and were excluded from the nucleus. Analysis of developing ectodermal and neural stem cells in MWCNTs injected embryos revealed normal division patterns and differentiation capacity. However, an increase in cell death of ectodermal but not of neural stem cells was observed, indicating stem cell-specific vulnerability to MWCNT exposure. The ease of CNT embryo injections, the possibility of detailed morphological and genomic analysis and the low costs make Drosophila embryos a system of choice to assess potential developmental and cellular effects of CNTs and test their use in future CNT based new therapies including drug delivery.

INTRODUCTION

The first report of the synthesis of carbon nanotubes (CNTs) two decades ago [1] sparked interest in such diverse fields as electronics, optics, physics, material sciences, medicine and biology. The promise CNTs hold for these fields originates from their unique physical, chemical, electrical and mechanical properties [2]. Consequently, commercial production and applications are increasing and CNTS have a growing presence in our daily lives ([3], see also Woodrow Wilson Nano Inventory). Accumulation of nanoparticles in our environment is still at the detection threshold but the continuous release of particles by production, wear and tear, and waste disposal makes an increased environmental exposure inevitable [4]. In addition, the future use of CNTs in medical applications such as drug delivery, biosensors and surgical scaffolds [5] will increase human contact with CNTs and justifies international efforts for the development and standardization of existing toxicity tests, as well as of new approaches to test the health impact of CNTs [6].

Environmental concerns and the hazard to human health associated with CNTs have attracted widespread attention [7], [8]. CNTs can cause cellular and tissue damage by stimulating inflammation and necrosis due to increased production of reactive oxygen species (ROS) [7], [9]. Single walled CNTs tend to be more damaging than multi-walled CNTs (MWCNTs) [10]. The shape, length and the addition of side groups also influence CNT toxicity [7], An increasing numbers of studies indicate that many of the toxic effects initially reported may be caused by contaminations deposited during CNT production, an observation explaining some of the inconsistencies in previous studies [7], [9]. Cell cultures are often the medium of choice for toxicity tests since they offer a fast, low cost and high-throughput approach. Yet, cell culture results vary with cell type and culture conditions [9], and results may not translate directly into the whole organism environment where, in a temporal and spatially controlled fashion, thousands of endogenous proteins and hundreds of different cell types interact with each other. Due to high costs, high throughput toxicity studies on mammals are scarce. It may be advantageous to opt for an alternative way, conducting high throughput studies in lower vertebrates and invertebrates with short generation time and high fecundity, and validate results obtained in these studies in a limited number of rodents. Indeed, zebrafish [11], [12], [13] and the flatworm C.elegans [14], [15] have been recently used to study the toxicity of CNTs. Both model organisms allow the establishment of basic mechanisms of CNT toxicity by examining viability, fertility, tissue and cellular integrity [16]. They also give an insight into alterations in gene expression changes, which underlie altered organ function [11], [15].

Here we present a third simple animal model system towards the study of CNT toxicity, Drosophila embryos. Embryos of Drosophila do not only offer all the advantages of zebrafish and C.elegans but also have unique features, which allow an unprecedented insight into the mechanisms of cellular toxicity. First, Drosophila is widely used to understand the biology of human diseases and also as a tool for gene and drug discovery [17]. Second, during early Drosophila embryogenesis, all nuclei share the same cytoplasm, which allows CNTs injected into the egg to interact with nuclei and to become included into the cells forming about 3 hours into embryogenesis. Injections overcome the problem of reduced bioavailability of CNTs and permit the study of CNT effects on cell division, cell survival, organ formation and function after cellular uptake, a likely occurrence during drug delivery or scaffold insertions upon surgery in humans [5]. Third, a plethora of available transgenic flies in which the expression of fluorescently tagged proteins outlines cellular substructures or distinct cell types can be used to study intra- and extracellular distribution of CNTs in living developing embryos [17], [18]. Finally, embryogenesis in Drosophila is a well-documented process, right down to the single cell level [19], [20], [21]. Hence any disturbances caused by the presence of CNT in cell movements or cell communication essential for organ formation or cellular differentiation can be easily detected.

In this report we used different subcellular markers to follow the distribution of MWCNTs in living, developing Drosophila embryos at a single cell resolution. We show that the intra- and extracellular accumulation of MWCNTs does not interfere with nuclear or cellular divisions or the overall embryonic development. Interestingly also the amount of DNA double-strand breaks known to contribute to genotoxic stress and cancer [22] are not significantly increased, However, MWCNTs induced a decrease in the survival of ectodermal stem cells, which was not observed in neural stem cells, suggesting that stem cell types differ in their vulnerability to MWCNTs exposure. Our study indicates that Drosophila can be used as a tool to study the toxicity of MWCNTs. The availability of near unlimited numbers of embryos, the ease of embryo injections and the low cost of the procedure make Drosophila embryos a model of choice to study developmental toxicity of CNTs in whole organisms.

RESULTS

Injected MWCNTs are incorporated into cells and do not disrupt embryonic development or cell motility

We sought to test the toxicity of MWCNTs for cell division, cellular differentiation and overall embryonic development by injecting MWCNTs into*Drosophila*embryos. In order to visualize live nuclear or cellular divisions, and subsequent organ formation, we used transgenic fly strains. either expressing histones coupled to Yellow Fluorescent Protein (YFP) labeling nuclei/chromosomes or Green Fluorescent Protein (GFP) trapped into the intron of the microtubule-binding protein Jupiter[23], labelling the outer cell membranes and mitotic spindles.

Drosophila embryogenesis starts with the syncytial blastoderm when embryos only consist of nuclei, which undergo 13 near synchronous divisions. Before the last 4 divisions, nuclei align along the embryo surface. After the last nuclear division, an actin-based movement results in the ingression of cell membranes from the egg membrane, and each nuclei and its surrounding cytoplasm is partitioned into a newly formed cell [24]. We injected MWCNTs and vehicle controls at the time when nuclei reach the embryonic periphery (Figure 1A). MWCNTs were marked with the red fluorescent lipophilic dye DiI (Invitrogen) as this dye has been shown to bind non-covalently to carbon nanotubes [25] and to be harmless for Drosophila embryogenesis [21]. MWCNTs were labelled at 1 mg/ml DiI in 100% DMSO. DiI in 100% DMSO forms a homogenous solution, which does not precipitate when spun at room temperature at 6000 rpm. However, addition of MWCNT results in the formation of a reddish-black precipitate. Microscopic inspection of this precipitate under epifluorescencevisualises small fluorescent puncta with the characteristic shape of MWCNTs. This observation concurs with previous reports showing binding of DiI to CNTs with high affinity [25]. MWCNT/DiI was injected as a colloidal suspension in 10% DMSO/water (see Experimental Section). As control we injected 100 μg/ml DiI in 50% DMSO to visualise the spread and accumulation of dye not bound to MWCNTs (Figure 1B). In contrast to 1 mg/ml DiI in DMSO, which diffused easily throughout the embryo labelling internal membranes (Figure 1B), injected MWCNTs formed small puncta, which only dispersed throughout four segments in the ventral half of the embryo (Figure 1C), equalling about 6% of total embryonic volume. The emission of unbound dye is no longer detectable after three nuclear divisions (data not shown) but the dye labelled MWCNT puncta can be followed throughout embryogenesis (Figure 1D–G) indicating that the dye remains bound to MWCNTs and does not diffuse away.

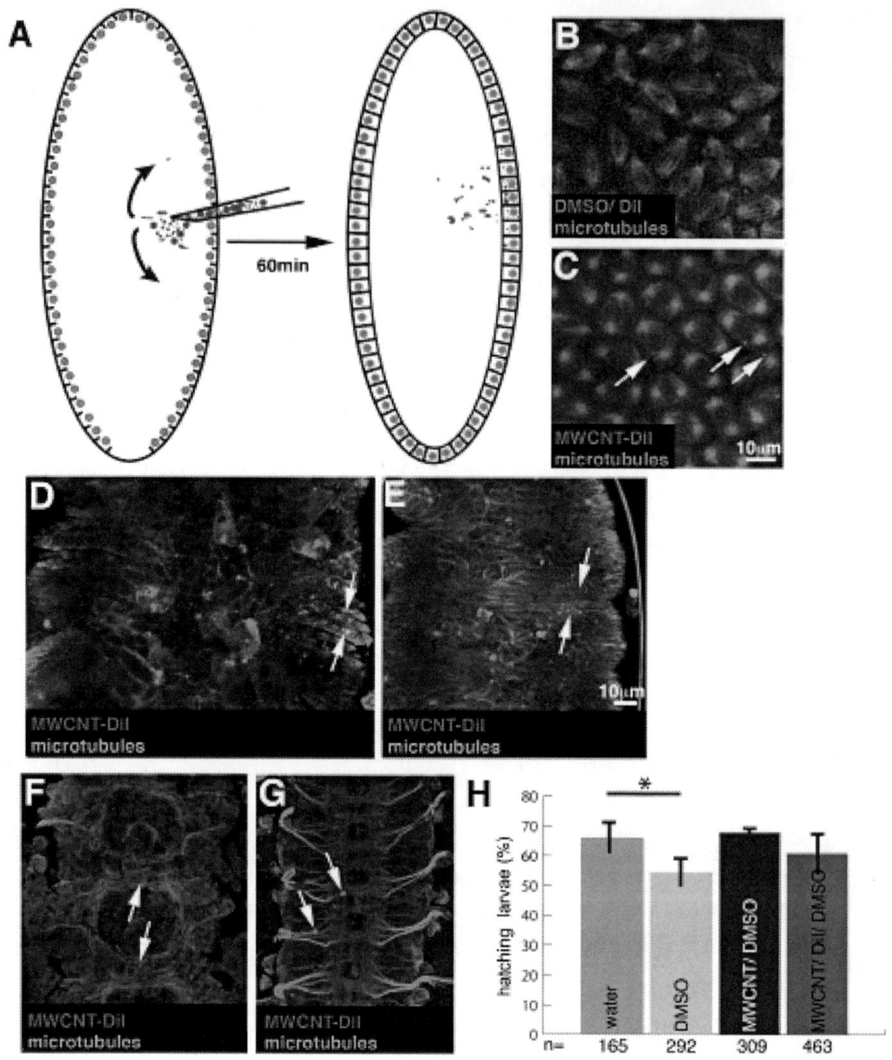

Figure 1. MWCNTs do not interfere with gross embryonic development and viability.

Ventral views; Anterior is up. Bar, 10 μm. (A) Using a microcapillary, we injected DiIlabelled MWCNTs (red) into the syncytial blastoderm of Drosophila embryos. At this stage of development, all nuclei (green) share the same cytoplasm permitting an unhindered diffusion (arrows) of the MWCNTs to nuclei adjacent to the injection site. In the next hour of embryogenesis, invaginating cell membranes will partition the nuclei into single cells incorporating the MWCNTs. (B, C) Live snapshots of embryos injected with 1 mg/ml DiI in DMSO (red, B, DMSO/DiI) or 1 mg/ml DiIlabelled MWCNTs in DMSO (red, C, MWCNT-DiI) immediately after injection into an embryo with microtubules labelled by GFP (green, GFP-jupiter). MWCNTs remain in the cytoplasm, the darker area surrounding the microtubules. Note that DiI not bound to MWCNT shows a diffuse membrane stain, whereas DiI bound to MWCNTs can be detected as small puncta (arrow). (D, E) Live confocal section of the same embryo 8 h (D) and 15 h (E) after injection. Cell outlines are labelled by GFP labelled microtubules (green, Jupiter-GFP). The majority of MWCNT (red) has been incorporated into the newly formed epidermis cells (arrows). (F, G) Live confocal section of the developing CNS of the same embryo 8 h (F) and 15 h (G) after injection. Cell outline and axons are labelled by GFP labelled microtubules (green, Jupiter-GFP). MWCNTs do not interfere with the initial outgrowth of axons (F, arrows) and are incorporated into the axonal scaffold without visible damage (G, arrows). (H) MWCNT injections do not interfere with embryonic viability. If embryos develop into healthy larvae, the larvae will hatch out of the egg shell. Hatching rate of embryos injected with water (blue, injection control), 10% DMSO (yellow, vehicle control), 1 mg/ml MWCNT in 10% DMSO/water (black, MWCNT/DMSO, see also Experimental Section) and 1 mg/ml MWCNT in 10% DMSO labelled with DiI (red, MWCNT/DiI/DMSO). Pairwise comparison with water injected embryos (t-Test), *, significant ($p<0.05$); error bars, StDev;. Y-axis, Hatching rate in % of injected embryos; X-axis, Number of injected embryos (n).

Injected MWCNTs stay mainly in the cytoplasm and rarely associate with microtubules (Movie S1). Even if MWCNTs collide with microtubules, microtubule based spindle formation proceeds normally (arrows, Movie S1). The presence of labelled or unlabelled MWCNTs does not interfere with the signalling process controlling the ingression of membranes during cellularisation [24]. In addition, MWCNT accumulation does not impede gastrulation (Movie S2), a highly coordinated movement of epithelial cells enforced by the apical constriction of actin [26]. Hence, MWCNT do not interfere with the formation and contraction of the two major motile fiber systems, microtubules and actin.

After cellularisation, MWCNTs are incorporated into ectodermal stem cells, which give rise to the epidermis (skin, Figure 1D, E) and into neural stem cells giving rise to neurons and glial cells, which become part of the central nervous system (CNS). Interestingly the presence of MWCNTs does not interfere with the extension and direction of neuronal axons (Figure 1F). At the end of embryogenesis, MWCNT have been incorporated into the CNS without causing any visible axonal damage (Figure 1G).

Drosophila embryos possess an innate immune system with a humoral component consisting of secreted anti-bacterial and anti-fungal peptides, and a cellular component made up of haematocytes that recognize and engulf foreign bodies and cell fragments [27]. Due to their increased size and their high motility, haematocytes can be easily detected and followed in the living embryo. During embryogenesis not all injected MWCNTs become incorporated in cells. Some MWCNT stay extracellular and together with MWCNT released from dying cells (see below) accumulate in older embryos. We recorded the behaviour of haematocytes around MWCNT deposits and surprisingly found that haematocytes do not specifically target these deposits (Movie S3). Although we detected some haematocytes with engulfed MWCNTs attached to cell debris (Movie S4), MWCNTs only seem to invoke a weak immune response in Drosophila embryos.

Finally, we tested the overall viability of embryos injected with MWCNTs and controls. At the end of development, healthy embryos will hatch as larvae. Hatching rate can therefore be used as readout for damage caused during embryogenesis. Hatching rate is never 100% because the preparation of embryos for injections always interferes with larval viability. Yet, as expected from our previous results the injection of either unlabeled or labeled MWCNTs did not reduce hatching rate as compared to water-injected controls (Figure 1H).

We conclude that the presence of intracellular MWCNTs has no observable effect on cell motility, cell communication, tissue and organ formation, phagocytosis or general viability of developing Drosophila embryos.

MWCNTs do not arrest DNA replication

The synchronised nuclear divisions in the syncytial blastoderm of Drosophila embryos offers the unique opportunity to check if the presence of MWCNTs interferes with DNA replication, and thereby slowing down or speeding up divisions. In addition, failure in DNA replication or chromosomal separation can easily be detected because these nuclei fall out of the embryonic surface into the yolk [28]. We injected MWCNTs and control vehicles into the syncytial blastoderm and recorded the last four synchronous nuclear divisions along the surface of the embryo (Movie S5, S6, S7). Nuclei surrounded by MWCNTs divided at the same time as adjacent MWCNT-free nuclei (Figure 2A, B). We did not detect any difference in the division cycles between control and MWCNT injected embryos or between injected and non-injected embryo halves (Figure 2C). We also did not observe any increase in nuclear fallout between embryos injected with MWCNTs or controls (data not shown). We conclude that MWCNTs do not interfere with DNA replication or chromosome separation. Interestingly, we never detected a nuclear incorporation of MWCNTs although the breakdown of the nuclear membrane during nuclear divisions would permit MWCNT entry. This argues for a yet to be described mechanism, which actively prevents the nuclear entry of MWCNTs.

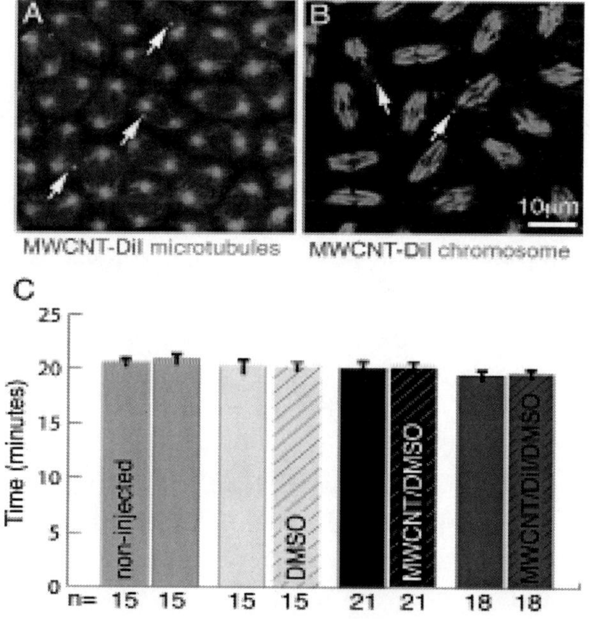

Figure 2. MWCNTs do not interfere with cell division.

Ventral views; Anterior is up. Bar equals 10 μm in A, B. (A) Single live confocal section through the syncytial blastoderm during a division wave. Microtubules are labelled with GFP (green, Jupiter-GFP). All nuclei are in the same stage (metaphase) of the cell cycle even if MWCNTs (red, arrows) are present. (B) Single live confocal section through an embryo with YFP labelled histones marking chromosomes (green). Regardless of the presence or absence of MWCNTs (red, arrows), all nuclei are in the same stage of division (anaphase). (C) The rate of nuclear divisions is not affected by the presence of MWCNTs. Division times for three nuclei per embryo half were recorded and division times between non-injected (non-hatched) and injected (hatched) halves compared. For non-injected embryos (grey) division times of three randomly chosen nuclei for each half were recorded. We detect no differences in the division times between non-injected and injected halves nor between embryos injected with 10% DMSO (yellow, solvent control), 1 mg/ml MWCNT in 10% DMSO/water (black, MWCNT/DMSO) and 1 mg/ml MWCNT in 10% DMSO labelled with DiI (red, MWCNT/DiI/DMSO). Y-axis, division time in minutes; X-axis, Number of nuclei analysed (n).

MWCNTs do not induce DNA double-strand breaks

A complex DNA repair system in eukaryotes ensures that during cell divisons chromosomes are segregated faithfully, and the stability of the genome and its information is preserved [29]. Endogenous recombinations and a multitude of external agents can cause genome instability resulting in mutations, genome rearrangements, chromosome fragmentation and chromosome loss. Genome instability is the major cause of cell death, aging and cancer [22]. A major trigger of genome instability is the unfaithful repair of DNA double-strand breaks. DNA double-strand breaks are marked for repair by a phosphorylated variant of the DNA binding molecule Histone2, γH2Av [30]. We used an antibody [31] to detect γH2Av after injection of MWCNTs (Figure 3). We injected labeled and unlabeled MWCNTs into the syncytial blastoderm along with vehicle controls (10% DMSO) and a positive control, Camptothecin, a substance known to cause DNA double-strand breaks [32]. Nine hours after injection the embryos were fixed and immunostained for γH2Av. The nuclei of the injected embryos were marked by the expression of YFP-Histone 2A (Gal4V2h/UAS::YFP::Histone 2A). We scored all γH2Av positive signal spots colocalising with the nucleus as double-strand breaks (Figure 3A–E′). Only the 10 truncal segments of the nervous system which originate from the injection site were analysed. As expected, we detected a highly significant ($p<0.001$) increase in the number of DNA double-strand breaks between non-injected embryos and Camptothecin injected embryos (Figure 3F). We also detected a significant ($P<0.05$) increase of double-strand breaks in embryos injected with 10% DMSO and DiIlabelled MWCNTs. Yet, these increases seem to be due to DiI or DMSO since compared to non-injected embryos, the injection of unlabeled MWCNTs does not increase the frequency of DNA double-strand breaks (Figure 3F). We conclude that the intracellular presence of MWCNT does not cause DNA double-strand breaks and hence does not significantly increase genotoxic stress.

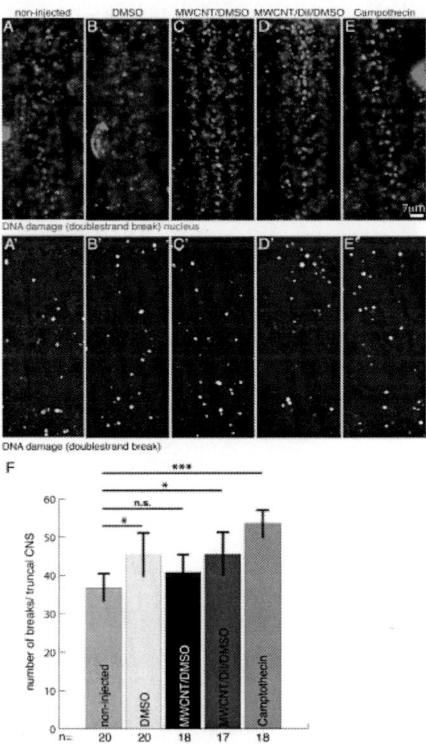

Figure 3. MWCNTs do not cause a significant increase in DNA double-strand breaks.

Ventral views; Anterior is up. Bar equals 7 μm in A–E′. Embryos were allowed to develop for 8 h after MWCNTs injection before fixation and immunohistochemistry. (A–E′) Immunostainings of double-strand breaks (red; anti-γH2Av) and nuclei (green; YFP-HistoneH2a, anti-GFP) in non-injected embryos and embryos injected with 10% DMSO (DMS0), 1 mg/ml MWCNT in 10% DMSO/water (MWCNT/DMSO), 1 mg/ml MWCNT in 10% DMSO labelled with DiI (MWCNT/DiI/DMSO) and Camptothecin (media indicated on top of each image). A′–E′ shows red channel only in grey. (F) The presence of MWCNT does not increase the frequency of DNA double-strand breaks.

To assess the number of double-strand breaks we counted all γH2Av puncta located in nuclei of the truncal CNS (1st thoracic to 7th abdominal segment). Camptothecin, an agent known to cause double-strand breaks, served as a positive control. The only highly significant increase in the frequency of breaks can be observed between non-injected (grey) and Camptothecin (green) injected embryos. We detect a significant increase between non-injected embryos and embryos injected with 10% DMSO (yellow, DMSO) or 1 mg/ml MWCNT labelled with DiI in 10% DMSO (red, MWCNT/DiI/DMSO). Yet, injection of 1 mg/ml MWCNT in 10% DMSO/water (black, MWCNT/DMSO) does not cause more double-strand breaks (pairwise comparison with non-injected embryos; t-Test). n.s., non significant; *, significant ($p<0.05$); ***, highly significant ($p<0.001$).Y-axis, number of breaks/truncal CNS; X-axis, Number of embryos analysed (n).

MWCNTs increase cell death of ectodermal Stem Cells

Injections of MWCNTs into the ventral region of Drosophila embryos allow following the influence of CNTs on survival, division pattern and progeny differentiation of two kinds of stem cells. Ectodermal stem cells, which give rise to the outer body cover, the larval epidermis and some cells of the sensory nervous system, and neural stem cells, which generate the neurons and glial cells in the ventral part of the central nervous system, the equivalent to the spinal cord in vertebrates [33], [34]. We injected MWCNTs and vehicle control (10%DMSO in water) into the syncytial blastoderm. 1.5 h after injection we labelled three to six cells at the injection site with the lipophilic dye DiI [21]. The dye is transferred from the stem cell to the progeny and labels the cell membrane of all progeny (Figure 4). We allowed the embryos to complete their development before we processed them for immunostaining and recorded the progeny. Consistent with our previous results (Figure 1), immunostaining against axonal or neuronal surface proteins did not show any gross morphological disturbances of the nervous sytem. This result is supported by the observed normal differentiation of epidermal and neural progeny derived from the labelled stem cells (Figure 4A–F). Ectodermal stem cells in non-injected embryos can give rise to 2–12 progeny with an average of 6.07 (+/−2.82, n = 202). In embryos injected with 10% DMSO progeny number varies between 2–11, average 6.04 (+/−2.56, n = 25) and injection of MWCNTs result in 2–12 daughter cells/ectodermal precursor, average 6.27 (+/−2.78, n = 48). The number of progeny derived from neural stem cells is stem cell specific [34] and we did not detect any differences between identical stem cells labelled in non-injected, 10% DMSO injected, and MWCNT injected embryos. We conclude that the presence of MWCNTs does not affect the number of progeny i.e. the division pattern of precursors cells.

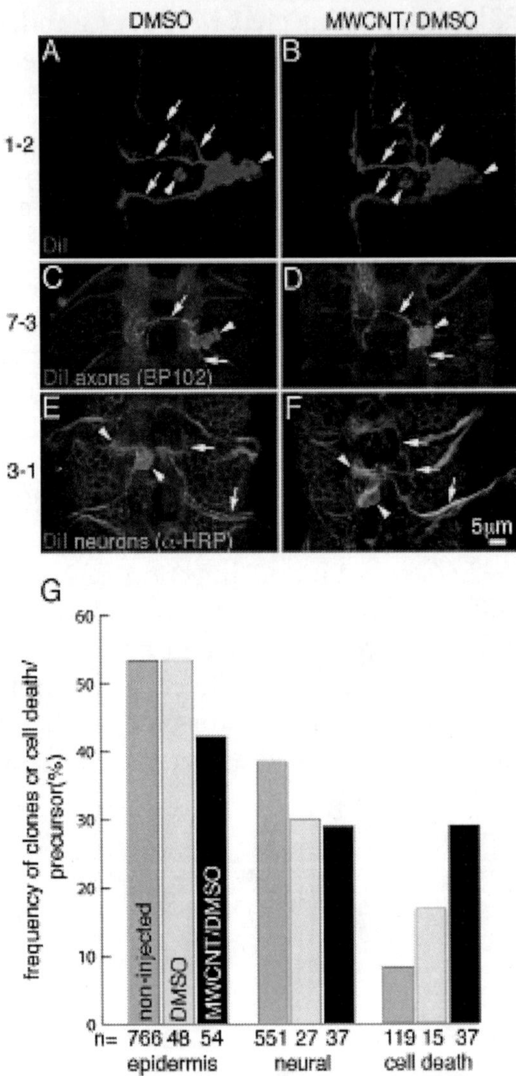

Figure 4. MWCNTs do not interfere with differentiation but increase cell death of ectodermal stem cells.

Ventral views; Anterior is up. Bar equals 5 μm in A–F. Single neural stem cells were labelled with DiI (red) in embryos injected with 10% DMSO (DMSO; A, C, E) or with 1 mg/ml MWCNT in 10% DMSO/ water (MWCNT/DMSO; B, D, F). The identification of the stem cell (1-2, 7-3, 3-1) is given on the left. (A–F) MWCNT does not interfere with neuronal differentiation. Labelled stem cells were allowed to divide and the progeny were examined for axonal extension (arrow), cell body position (arrowhead) and cell number. To examine the overall development of the CNS either all axons (anti-BP102; C, D) or all neuronal membranes, axons and cell body (anti-HRP; E, F) were immunostained. We do not detect any differences between stem cell progeny in non-injected embryos [34], DMSO or MWCNT/DMSO injected embryos. Note that the difference in branching pattern of 3-1 progeny (E, F) is caused by their different anterior-posterior location [34] and not MWCNT injections. (G) Frequency and kind of progeny obtained from labelled embryonic stem cells. Compared to non-injected embryos (grey), the injection of DMSO (yellow) results in an increase of cell death of neural stem cells. Injection of MWCNT/ DMSO (black) affects the survival of epidermal as well as neural stem cells resulting in a significant ($p<0.05$) increase in cell death. The cell death increase observed in the case of neural stem cells is not higher than upon injection of DMSO only. Y-axis, frequency in % of epidermal (epidermis; column 1–3) or neural progeny (neural; column 4–6) or cell death (column 7–9) arisen from labelled stem cells; X-axis, total number of cells labelled (n). Numbers for non-injected embryo are taken from Bossing et al., 1996.

If labeled stem cells or progeny die we detect at the end of embryogenesis labeled cell debris in haematocytes accompanied by a reduced number or a complete loss of progeny. In 10% DMSO injected embryos we noticed an increased cell death of neural stem cells but not ectodermal stem cells. In MWCNT injected embryos, cell death of neural stem cells is as high as in 10% DMSO injected controls, arguing that DMSO is the insulting agent. In contrast, cell death of ectodermal stem cells in MWCNT injected embryos was higher than in embryos injected with 10% DMSO alone (Figure 4G). We conclude that the intracellular presence of MWCNTs reduces viability of distinct types of stem cells.

DISCUSSION

The increased production and application of CNTs makes the testing of their environmental and health impact an urgent necessity. The ingestion of CNTs by model organisms such as mice, Drosophila, C. elegans and zebrafish is a convenient way to study the impact of CNTs released into the environment (e.g. [12], [15], [35], [36]).

Yet, ingested CNTs mainly reside in the external digestive system and their cellular uptake is limited [37]. Future medical applications of CNTs will cause permanent exposure of internal organs, e.g. by the implantation of CNT coated stents or scaffolds. Wear of these scaffolds will release CNT particles and promote the subsequent cellular uptake by phagocytic cells of the immune and digestive system. In the case of drug delivery systems to combat cancer or promote stem cell proliferation, uptake into dividing cells is the ultimate objective [5]. We show that injection of CNTs into early Drosophila embryos can serve as a cost effective, convenient and informative testing system to study the biological effects of cellular uptake of MWCNTs.

We labeled MWCNTs by incubation with the lipophilic dye DiI dissolved in DMSO. In agreement with previous results [25], microscopical inspection under epifluorescence of the precipitate after several washes with DMSO confirms the binding of the dye to MWCNTs. Injected DiI dissolved in DMSO spreads through cellular membranes and rapidly becomes diluted during cell divisions. In contrast, injected MWCNT/DiI/DMSO solutions do not show any diffusion of the dye into membranes but form puncta inside the cytoplasm. These puncta can be visualized throughout embryogenesis, indicating that the dye stays bound to MWCNTs after injection into embryos.

We show that injection of labeled and unlabeled MWCNTs into the syncytial blastoderm of the embryo does not interfere with nuclear divisions and results in the cellular uptake of MWCNTs when nuclei become enclosed by ingrowing cell membranes. Using transgenic embryos with cellular substructures labeled by fluorescent proteins, we detect no accumulation of intracellular DiIlabelled MWCNTs along microtubules, cell membranes or inside nuclei. Intracellular MWCNTs are dispersed throughout the cytoplasm and randomly distributed during cell divisions.

Organ formation during embryogenesis relies on cell communication achieved by an intricate network of molecular signaling pathways [38]. The extracellular and intracellular presence of MWCNT seems not to disrupt organ formation and hence does not reduce embryonic viability. Our single cell analysis of the developing CNS, the most complex organ, demonstrates that MWCNTs do not alter stem cell divisions or neuronal and glial differentiation. Surprisingly, axon outgrowth and guidance, processes controlled by various signaling pathways [39], [40], proceed normally in the presence of MWCNTs, which even become incorporated into the mature larval CNS without any visible disturbances.

Two types of stem cells originate from the injected embryonic ventral region, ectodermal stem cells, giving rise to the epidermis and sensory nervous system of the mature larvae, and neural stem cells giving rise to neurons and glia cells. All the progeny of both stem cell types have been identified [34], [41]. The development and division pattern of these stem cells is tightly controlled. For example, following the loss of function of tumour suppressor genes, an increased number of undifferentiated progeny from neural stem cells is observed [42]. Yet, the presence of MWCNTs does not interfere with cell number or differentiation of neural progeny. In addition, MWCNT injections do not increase the frequency of DNA double-strand breaks, which are a major cause of genotoxicstress and carcinogenesis [22].

Surprisingly, we detect that the type of stem cell defines the risk of damage caused by the cellular uptake of MWCNTs. Ectodermal stem cells exhibit increased cell death but not neural stem cells. Both stem cell types originate from the same cell layer by cell-cell interactions from a cluster of about 4–6 cells which are morphologically and genetically equivalent [43], [44]. One cell of this cluster will become a neural stem cell and the remaining cells adopt the ectodermal stem cell fate. At the time of the first wave of selection, shortly after gastrulation, we do not detect an accumulation of MWCNTs in only one cell but a spread over adjacent cells (Movie S2). It is unlikely that minor differences in MWCNT intracellular load increases the possibility to become an ectodermal stem cell and therefore increases the number of ectodermal stem cell death. Such an imbalance in cell fate would also lead to a reduced number of labelled neural stem cells, which we do not observe. Immediately after their selection both stem cell types become different from each other not only in their mode and number of divisions but also in their molecular identity. Ectodermal stem cells stay in the superficial tissue they originated in and divide symmetrically and equally, whereas neural stem cells delaminate to form a new tissue below the ectodermal layer and divide asymmetrically and unequally [33].

Ectodermal stem cells give rise to a maximum of 12 progeny but neural stem cells can give rise to up to 40 progeny [34]. If number of divisions and hence dilution of the MWCNT load would be crucial for stem cell survival, we would expect that early born progeny which inherit a larger load of MWCNT are less viable than later born progeny resulting in a reduction of the average number of progeny per stem cell. However, the average number of progeny per stem cell in injected and non-injected embryos is not different. Hence, the reduced number of divisions in ectodermal stem cells seems not to be the cause for the increased cell death. Future studies using for example current sequencing techniques allowing the rapid identification of transcripts differentially expressed in both stem cell types, will aid the identification of the pathways responsible for the increased susceptibility to MWCNT caused damage in ectodermal stem cells versus neural stem cells.

We demonstrate that Drosophila embryos can be used as a model to test the toxicity of CNTs in unprecedented detail. Drosophila embryos are an excellent system to study developmental toxicity because of easy and inexpensive maintenance, ready availability and the possibility of automating embryo injections. Can results obtained from injections into embryos be used to evaluate toxicity of CNTs for other animals and even humans? Drosophila has a long-standing history as genetic model for human diseases [45]. All signaling pathways and most of the organs in humans have analogous pathways or organs in Drosophila, which is also expressed in the high conservation rate of 55% of all genes between the Drosophila and human genomes [45], [46]. Basic cell biological aspects such as cell division including oncogenesis, transcriptional regulation, membrane biology, DNA integrity and repair, cell movement and tissue formation are conserved all over the animal kingdom, and often have been first studied in Drosophila. These arguments underline the strength of Drosophila embryos as a novel model for future studies of CNT developmental toxicity.

MATERIALS AND METHODS

Fly strains

We used the GAL4 driver, GAL4^{V2h}[47], the protein trap Jupiter-GFP[23]and UAS-YFP::HistoneH2[48].

Carbon nanotube labeling and injections

MWCNT were supplied by NanoAmor Europe (Cat.No. 1233YJ, ID: 5–15 nm, OD: 50–80 nm, Length 10–20 μm; analysis by Energy Dispersive X-ray Spectroscopy: C: 97.37%, Cl: 0.2%, Fe: 0.5%, Ni: 1.86%, S: 0.02%). We dissolved 10 mg MWCNTs in 1 ml 100% DMSO (Sigma, Cat. No. D8418, 99.9% pure, impurities: <0.1 water, <0.001 meq/g titratable acid) with or without the addition of 1 mg of the lipophilic dye, DiI (Cat. No. D-282, Invitrogen). Both solutions were vortexed on high speed for about 2 min to coat MWCNTs with DMSO or DMSO/DiI. DiIlabelled MWCNTs were washed five times with 100% DMSO to remove excess, unbound dye. After final wash, MWCNTs were spun down with a benchtop centrifuge at 6000 rpm and checked for fluorescence [20× objective, Numerical aperture (N.A.): 0.75, 10× ocular, excitation: 510–560 nm, dichroic: 565 nm, Barrier: 590 nm], Stocks were kept at 4°C in the dark. DiI label of MWCNTs is stable for about 6 weeks. For injections we warmed the stock solution and de-ionised water at 65°C for 10 min, added 10 ul of the stock solution into 90 ul of water and vortexed the solution for 5 min at medium speed. After vortexing, MWCNTs form a transient colloidal dispersion, which has to be injected immediately. Before injection, CNTs in the solution was examined for dispersion and epifluorescence under 200× magnification (Nikon TE 300, 20× objective, N.A. 0.75, 10× ocular). We injected MWCNTs using borosilicate capillaries (Harvard Apparatus; ID: 0.75 mm, OD: 1.5 mm, no inner filament) pulled on a Sutter P-97 micropipette puller and with a tip bevelled to an angle of 25° using a BachoferMicrotip Grinder. Embryos were de-chorionated manually on double-sided tape, aligned on a block of 2.5% Agar and transferred with their ventral side down onto a coverslip (24 mm×60 mm) of which the middle was coated with glue and to which a plastic frame was attached (cut out from self adhesive book binding foil, ID: 16 mm×16 mm, OD: 22 mm×22 mm). The coverslip was supported with a microscope slide, embryos were desicated at room temperature for 5–10 min (depending on humidity) and covered with halocarbon oil (Voltalef 10 s) to stop further evaporation. Injection of embryos was controlled with 100× magnification (10× objective, N.A.0.45, 10× ocular), which allows monitoring of dispersion of MWCNTs in the capillary. On average about 20 pl of the 1 mg/ml solution, equalling about 4% of total embryo volume, was injected equalling 5 pg of MWCNTs/ embryo. MWCNTs only showed a limited diffusion covering about 6% of total embryo volume. Embryos injected as control with water, 10% DMSO, unlabelled MWCNTs or with Camptothecin were treated the same way.

Immunohistochemistry

The following antibodies were used BP102 (mouse, 1:1000; kindly provided by Professor Nipam Patel), anti- γH2Av (1:1000; kindly provided by Kim McKim) and anti-HRP coupled to FITC (goat, 1:500, MPbio). Embryos were stained as fillets as previously described[34]. All antibodies are used for routine stains in*Drosophila*embryonic research and their staining pattern has been well described[49][31].

Time lapse, Confocal and Image processing

Fixed and live samples were recorded using a Zeiss LSM710. For time lapse recording of living embryos, recording started 15 min after injection and intervals were set to 30 sec. A maximum of three Z-levels covering 2 μm were scanned bi-directionally for each time point. Movies were generated using the ZEN software (Zeiss) and Image J. Movies and Images were assembled and labelled using Photoshop CS6.

ACKNOWLEDGMENTS

The authors would like to thank C. Barros, R. Handy, I. Robinson and R. Fern for critical reading of the manuscript.

AUTHOR CONTRIBUTIONS

Conceived and designed the experiments: TB EMC. Performed the experiments: BL TB. Analyzed the data: TB. Contributed reagents/materials/analysis tools: TB.Wrote the paper: TB. Critical review of the manuscript: BL EMC.

REFERENCES

1. Iijima S (1991) Helical microtubules of graphitic carbon. Nature 354: 56–58. doi: 10.1038/354056a0
2. Dresselhaus MS, Dresselhaus G, Charlier JC, Hernandez E (2004) Electronic, thermal and mechanical properties of carbon nanotubes. Philos Trans A Math PhysEngSci 362: 2065–2098.doi: 10.1098/rsta.2004.1430
3. De Volder MF, Tawfick SH, Baughman RH, Hart AJ (2013) Carbon nanotubes: present and future commercial applications. Science 339: 535–539. doi: 10.1126/science.1222453
4. Gottschalk F, Sun T, Nowack B (2013) Environmental concentrations of engineered nanomaterials: Review of modeling and analytical studies. Environ Pollut 181: 287–300. doi: 10.1016/j.envpol.2013.06.003
5. Thorley AJ, Tetley TD (2013) New perspectives in nanomedicine. PharmacolTherdoi: 10.1016/j.pharmthera.2013.06.008
6. Juillerat-Jeanneret L, Dusinska M, Fjellsbo LM, Collins AR, Handy RD, et al. (2013) Biological impact assessment of nanomaterial used in nanomedicine. Introduction to the NanoTEST project. Nanotoxicologydoi: 10.3109/17435390.2013.826743
7. Liu Y, Zhao Y, Sun B, Chen C (2012) Understanding the Toxicity of Carbon Nanotubes. AccChem Res doi: 10.1021/ar300028m
8. Lanone S, Andujar P, Kermanizadeh A, Boczkowski J (2013) Determinants of carbon nanotube toxicity. Adv Drug Deliv Rev doi: 10.1016/j.addr.2013.07.019
9. Lewinski N, Colvin V, Drezek R (2008) Cytotoxicity of nanoparticles. Small 4: 26–49. doi: 10.1002/smll.200700595
10. Jia G, Wang H, Yan L, Wang X, Pei R, et al. (2005) Cytotoxicity of carbon nanomaterials: single-wall nanotube, multi-wall nanotube, and fullerene. Environ SciTechnol 39: 1378–1383. doi: 10.1021/es048729l
11. da Rocha AM, Ferreira JR, Barros DM, Pereira TC, Bogo MR, et al. (2013) Gene expression and biochemical responses in brain of

zebrafishDaniorerio exposed to organic nanomaterials: carbon nanotubes (SWCNT) and fullerenol (C60(OH)18-22(OK4)). Comp BiochemPhysiol A MolIntegrPhysiol 165: 460–467. doi: 10.1016/j.cbpa.2013.03.025

12. Cheng J, Cheng SH (2012) Influence of carbon nanotube length on toxicity to zebrafish embryos. Int J Nanomedicine 7: 3731–3739. doi: 10.2147/ijn.s30459
13. Shaw BJ, Handy RD (2011) Physiological effects of nanoparticles on fish: a comparison of nanometals versus metal ions. Environ Int 37: 1083–1097. doi: 10.1016/j.envint.2011.03.009
14. Nouara A, Wu Q, Li Y, Tang M, Wang H, et al. (2013) Carboxylic acid functionalization prevents the translocation of multi-walled carbon nanotubes at predicted environmentally relevant concentrations into targeted organs of nematode Caenorhabditiselegans. Nanoscale 5: 6088–6096. doi: 10.1039/c3nr00847a
15. Chen PH, Hsiao KM, Chou CC (2013) Molecular characterization of toxicity mechanism of single-walled carbon nanotubes. Biomaterials 34: 5661–5669. doi: 10.1016/j.biomaterials.2013.03.093
16. Handy RD, Cornelis G, Fernandes T, Tsyusko O, Decho A, et al. (2011) Ecotoxicity test methods for engineered nanomaterials: practical experiences and recommendations from the bench. Environ ToxicolChem 31: 15–31. doi: 10.1002/etc.706
17. Kasai Y, Cagan R (2012) Drosophila as a tool for personalized medicine: a primer. Per Med 7: 621–632.doi: 10.2217/pme.10.65
18. Manning L, Heckscher ES, Purice MD, Roberts J, Bennett AL, et al. (2012) A resource for manipulating gene expression and analyzing cis-regulatory modules in the Drosophila CNS. Cell Rep 2: 1002–1013. doi: 10.1016/j.celrep.2012.09.009
19. Nusslein-Volhard C, Wieschaus E (1980) Mutations affecting segment number and polarity in Drosophila. Nature 287: 795–801. doi: 10.1038/287795a0
20. Campos-Ortega J, Hartenstein V (1997) The embryonic development of Drosophila melanogaster. New York Berlin: Springer.
21. Bossing T, Technau GM (1994) The fate of the CNS midline progenitors in Drosophila as revealed by a new method for single cell labelling. Development 120: 1895–1906.
22. View Article PubMed/NCBI Google Scholar
23. Aguilera A, Garcia-Muse T (2013) Causes of Genome Instability. Annu Rev Genet doi: 10.1146/annurev-genet-111212-133232
24. Morin X, Daneman R, Zavortink M, Chia W (2001) A protein

trap strategy to detect GFP-tagged proteins expressed from their endogenous loci in Drosophila. ProcNatlAcadSci U S A 98: 15050–15055. doi: 10.1073/pnas.261408198

25. Mazumdar A, Mazumdar M (2002) How one becomes many: blastodermcellularization in Drosophila melanogaster. Bioessays 24: 1012–1022. doi: 10.1002/bies.10184

26. Prakash R, Washburn S, Superfine R, Cheney RE, Falvo MR (2003) Visualisation of individual carbon nanotubes with fluorescent microscopy using conventional fluorophores. Applied Physics Letters 83: 1219–1221. doi: 10.1063/1.1599042

27. Sawyer JM, Harrell JR, Shemer G, Sullivan-Brown J, Roh-Johnson M, et al. (2010) Apical constriction: a cell shape change that can drive morphogenesis. DevBiol 341: 5–19. doi: 10.1016/j.ydbio.2009.09.009

28. Kounatidis I, Ligoxygakis P (2012) Drosophila as a model system to unravel the layers of innate immunity to infection. Open Biol 2: 120075. doi: 10.1098/rsob.120075

29. Frenz LM, Glover DM (1996) A maternal requirement for glutamine synthetase I for the mitotic cycles of syncytial Drosophila embryos. J Cell Sci 109 (Pt 11) 2649–2660.

30. Iyama T, Wilson DM 3rd (2013) DNA repair mechanisms in dividing and non-dividing cells. DNA Repair (Amst) 12: 620–636. doi: 10.1016/j.dnarep.2013.04.015

31. Talbert PB, Henikoff S Histone variants–ancient wrap artists of the epigenome. Nat Rev Mol Cell Biol 11: 264–275. doi: 10.1038/nrm2861

32. Mehrotra S, McKim KS (2006) Temporal analysis of meiotic DNA double-strand break formation and repair in Drosophila females. PLoS Genet 2: e200. doi: 10.1371/journal.pgen.0020200

33. Ryan AJ, Squires S, Strutt HL, Johnson RT (1991) Camptothecin cytotoxicity in mammalian cells is associated with the induction of persistent double strand breaks in replicating DNA. Nucleic Acids Res 19: 3295–3300. doi: 10.1093/nar/19.12.3295

34. Technau GM (1987) A single cell approach to problems of cell lineage and commitment during embryogenesis of Drosophila melanogaster. Development 100: 1–12.

35. Bossing T, Udolph G, Doe CQ, Technau GM (1996) The embryonic central nervous system lineages of Drosophila melanogaster. I. Neuroblast lineages derived from the ventral half of the neuroectoderm. DevBiol 179: 41–64. doi: 10.1006/dbio.1996.0240

36. Cheng J, Chan CM, Veca LM, Poon WL, Chan PK, et al. (2009) Acute and long-term effects after single loading of functionalized multi-walled

carbon nanotubes into zebrafish (Daniorerio). ToxicolApplPharmacol 235: 216–225. doi: 10.1016/j.taap.2008.12.006

37. Philbrook NA, Walker VK, Afrooz AR, Saleh NB, Winn LM (2011) Investigating the effects of functionalized carbon nanotubes on reproduction and development in Drosophila melanogaster and CD-1 mice. ReprodToxicol 32: 442–448. doi: 10.1016/j.reprotox.2011.09.002
38. Galloway T, Lewis C, Dolciotti I, Johnston BD, Moger J, et al. (2010) Sublethal toxicity of nano-titanium dioxide and carbon nanotubes in a sediment dwelling marine polychaete. Environ Pollut 158: 1748–1755. doi: 10.1016/j.envpol.2009.11.013
39. Gilbert SF (2010) Developmental Biology; Sunderland, editor: Sinauer Associates.
40. Tessier-Lavigne M, Goodman CS (1996) The molecular biology of axon guidance. Science 274: 1123–1133. doi: 10.1126/science.274.5290.1123
41. Furrer MP, Chiba A (2004) Molecular mechanisms for Drosophila neuronetwork formation. Neurosignals 13: 37–49. doi: 10.1159/000076157
42. Schmidt H, Rickert C, Bossing T, Vef O, Urban J, et al. (1997) The embryonic central nervous system lineages of Drosophila melanogaster. II. Neuroblast lineages derived from the dorsal part of the neuroectoderm. DevBiol 189: 186–204. doi: 10.1006/dbio.1997.8660
43. Choksi SP, Southall TD, Bossing T, Edoff K, de Wit E, et al. (2006) Prospero acts as a binary switch between self-renewal and differentiation in Drosophila neural stem cells. Dev Cell 11: 775–789. doi: 10.1016/j.devcel.2006.09.015
44. Campos-Ortega JA (1995) Genetic mechanisms of early neurogenesis in Drosophila melanogaster. MolNeurobiol 10: 75–89. doi: 10.1007/bf02740668
45. Skeath JB, Thor S (2003) Genetic control of Drosophila nerve cord development. CurrOpinNeurobiol 13: 8–15. doi: 10.1016/s0959-4388(03)00007-2
46. Bier E (2005) Drosophila, the golden bug, emerges as a tool for human genetics. Nat Rev Genet 6: 9–23. doi: 10.1038/nrg1503
47. Venter JC, Adams MD, Myers EW, Li PW, Mural RJ, et al. (2001) The sequence of the human genome. Science 291: 1304–1351. doi: 10.1016/s0002-9394(01)01077-7
48. Hacker U, Lin X, Perrimon N (1997) The Drosophila sugarless gene modulates Wingless signaling and encodes an enzyme involved in polysaccharide biosynthesis. Development 124: 3565–3573.

49. Bellaiche Y, Gho M, Kaltschmidt JA, Brand AH, Schweisguth F (2001) Frizzled regulates localization of cell-fate determinants and mitotic spindle rotation during asymmetric cell division. Nat Cell Biol 3: 50–57. doi: 10.1038/35050558

50. Bashaw GJ (2010) Visualizing axons in the Drosophila central nervous system using immunohistochemistry and immunofluorescence. Cold Spring HarbProtoc 2010: pdb prot5503. doi: 10.1101/pdb.prot5503

Chapter 12

ROLE OF TEMPERATURE AND CARBON NANOTUBE REINFORCEMENT ON EPOXY BASED NANOCOMPOSITES

Subhranshu Sekhar Samal

Centre for Nanoscience & Nanotechnology (A joint initiative of IGCAR, Kalpakkam & Sathyabama University, Chennai) India-600119

E-mail: sekhar_nitrkl@rediffmail.com, Cell: +91-9962369727

ABSTRACT

This paper presents the synthesis of epoxy based Multiwalled Carbon nanotubes (MWCNTs) reinforced composites by method of sonication. The variation in the nature of reinforcement (Aligned & Randomly oriented MWNTS) has resulted in the improvement of mechanical properties like flexural modulus, tensile strength and hardness. A small change in chemical treatment of the nanotubes has a great effect in the mechanical and morphological properties of nanocomposites due to effective load transfer mechanism and state of dispersion. The change in properties has been verified by optical microscopy and Scanning electron microscopy. Apart from that the prepared composites has been treated under different temperatures (like hot water, room temperature and liquid nitrogen temperature) and the change in mechanical as well as morphological nature has been verified by SEM of Fractographic surface this proved the elasticity and ductility of the composites.

INTRODUCTION

The discoveries of carbon nanotubes (CNT) have initiated researches in many different areas; one of the most intriguing applications of CNT is the polymer/CNT nanocomposites [1-5]. The high mechanical, electrical and thermal property of CNTs make them ideal candidate as fillers in lightweight polymer composite [6]. Due to their high specific strength and stiffness, nanotube-reinforced polymer composites have become attractive structural materials not only in the weight-sensitive aerospace industry, but also in the marine, armor, automobile, railway, civil engineering structures, and sporting goods industries. Epoxy resin is the polymer matrix used most often with reinforcing nanotubes for advanced composite applications. The resins of this class have good stiffness, specific strength, dimensional stability, and chemical resistance, and show considerable adhesion to the embedded fiber [7]. Because micro-scale fillers have successfully been synthesized with epoxy resin [8–11] nanoparticles, nanotubes, and nanofibers are now being tested as filler material to produce high performance composite structures with enhanced properties [12–14]. Scanning Electron Micrographs shows (Fig.1) shows the Nanotube structures with diameter.

EXPERIMENTAL

Materials

Multiwall carbon nanotubes (MWCNTs) used for the preparation of nanocomposites was obtained from MER Corporation, USA. They are produced by arc plasma method (purity 95%, length 10-50 μm and diameters 20-70 nm). SEM morphology of the products (Fig. 1) was carried out with a «JEOL JSM-6480 LV Scanning Microscope».

Aligned carbon nanotubes (ACNTs) used for the preparation of nanocomposites was obtained from ARCI, Hyderabad, India. They are produced by chemical vapor deposition method diameters 10-20 nm. SEM morphology of the products (Fig.1 c) was carried out with a "JEOL JSM 5410 Scanning Microscope».

Figure (1a, 1b & 1c). Scanning Electron Micrographs of Multiwall carbon nanotubes (MWCNTs) showing diameter Epoxy polymer matrix was prepared by mixing epoxy resin (Ciba-Geigy, araldite LY-556 based on Bisphenol A) and hardener HY-951 (aliphatic primary amine) in wt. ratio 100/12.

Epoxy resin (5.3-5.4 equiv/kg) was of low processing viscosity &overall good mechanical properties.

MWNT/Epoxy Composite Preparation

Nanocomposites were prepared by the method of sonication. To achieve better state of dispersion first the nanotubes were treated with alcoholic medium for the deagglomeration of the tube bundles. The treated tubes were then added to the epoxy resin and sonicated for 2 hrs at room temperature. Then the mixture was cured under vacuum at 900 C for 10 hrs and another set of samples were cured under refrigeration condition to get ductile samples. The prepared samples were treated at 800 C for 6 hrs in the oven to remove the moisture contents of the samples. The finally prepared samples were treated at different environmental conditions for 24 hrs before the mechanical tests were performed. Few samples were treated by hot water at 800 C and some are treated in the cryofreezer at liquid nitrogen temperature (- 800 C).

Nomenclature

The nomenclature of the samples is as given in Table.1.

Sample Identification	Nature of the Sample
0	Plane/Neat Epoxy (Brittle)
1	Epoxy/CNT Composite (Brittle)
D0	Plane/Neat Epoxy (Ductile)
D1	Epoxy/CNT Composite (Ductile)
H0	Plane Epoxy Hot water treated (Brittle)
H1	Epoxy/CNT Composite Hot water treated (Brittle)
DH0	Plane/Neat Epoxy Hot water treated (Ductile)
DH1	Epoxy/CNT Composite Hot water treated (Ductile)
C0	Plane Epoxy Cryofreezer treated (Brittle)
C1	Epoxy/CNT Composite Cryofreezer treated (Brittle)
DC0	Plane/Neat Epoxy Cryofreezer treated (Ductile)
DC1	Epoxy/CNT Composite ductile Cryofreezer treated (Ductile)

Flexural & Tensile Modulus Test

Four types of flexural test samples were fabricated: they were pure epoxy beam & its ductile one and the nanotube composite beam & its ductile one. The beams were placed into different temperature environments for 24 hours prior the test, these were: (a) room temperature (20 0C), (b) warm water (80 0C) and (c) liquid nitrogen (-180 0C). From each sample, five rectangular specimens were taken for three-point bend test as per ASTM D790 (width=2.7cm, thickness=0.7cm, span=11.2cm, length=12cm). Flexural tests were carried out at ambient temperature using Instron-1195 keeping the cross-head speed 2 mm/min. Flexural modulus of each sample was determined from the average value of five specimens. The samples were made in the form of dog bone shape as shown (Fig.2) and the testings were carried out carried out with INSTRON-1195.

Tensile tests were carried out at room temperature with a constant cross-head rate of 2 mm/min. The choice of this quite low loading rate is to consider the brittle character of composites .Ten specimens were tested for each sample and the average value was obtained from the data of these measured specimens (Fig.2). The thickness of samples is 3 mm ± 0.5 mm.

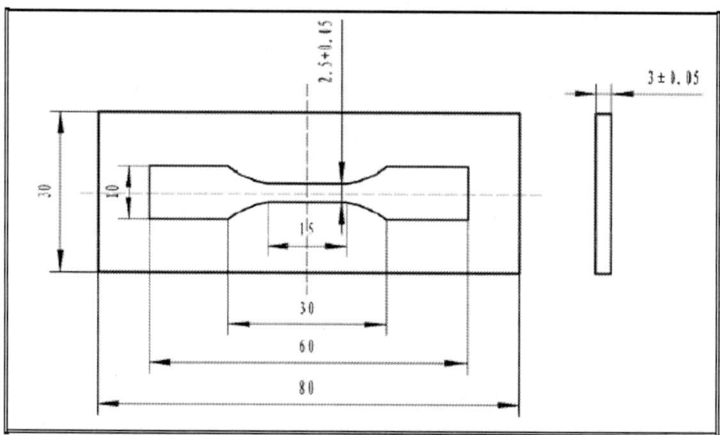

Figure 2: Dog Bone Shape sample for tensile modulus measurement.

Flexural measurements

Flexural modulus of pure resin, RCNT composite & ACNT composite are shown in Fig.7a. Both the composite samples are showing greater modulus than pure resin sample that is attributed to the high mechanical strength of Composite sample.

Fracture Surface Topography Characterization

Scanning electron microscope (JEOL-JSM-6480 LV) was used to conduct the fracture surface topography characterization. The cured samples were fractured and the fracture surfaces were coated with a thin platinum layer to study the morphology.

RESULTS AND DISCUSSION

Flexural Modulus, Tensile modulus & Young's Modulus Measurement

Flexural modulus of pure brittle resin (sample-0) & its respective CNT composite (sample-1) and ductile resin (sample-D0) & its CNT composite samples (D1) are found to be 24.52, 136.86, 44.15 &76.38 respectively. Increase in flexural modulus is more pronounced in samples treated in hot water for both brittle (H0-49 & H1-148.8) and ductile (DH0-64 & DH1-210) samples in comparison to samples treated in cryofreezer (C0-33,C1-140.2, DC0- 36, DC1-159.3) as shown in the Figure(3&4).

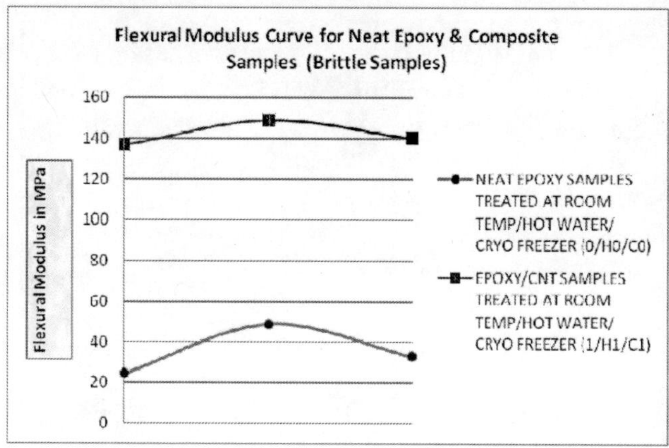

Figure 3. Flexural modulus of Neat epoxy & Composite samples treated at different conditions (Brittle Samples).

Because at high temperature the filler can constrain the mobility of polymer chains as well as their relaxation spectra [25], which can change the glass transition temperature [25, 26] and modulus of the matrix. All the composite samples are showing greater modulus than pure resin samples. This may be due to the high mechanical strength of CNT. The plane ductile sample after submerged into liquid nitrogen (DC0) revealed reduction in flexural modulus in comparison to room temperature treated sample (D0). The cause of reduction is may be due to the structural non-homogeneity and/or existence of a weak bonding interface between the nanotubes and surrounding matrix. The results of the refrigerated samples are showing variable behaviour. Ductile samples, which were, treated in hot water (HD0 & HD1), showed the best result. And those kept in cryofreezer became more brittle and hence less modulus. This may be due to contraction of the matrix, which increased the clamping stress to the nanotube surface, and thus increased the frictional force between the nanotubes and the matrix.

The general tendency observed from the result is that both the tensile strength and the fracture strain changed with variation of the thermal conditioning while Young's modulus (Elastic Modulus) reduced at the same time. The tensile strength exhibited a maximum value of 69.7 MPa (DH1) (Fig.5) and the flexural modulus reached the highest value of 210 MPa (DH1) too (Fig.4).

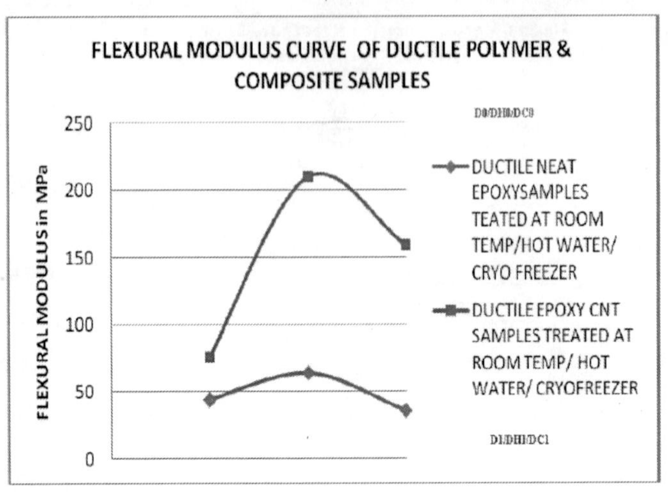

Figure 4. Flexural modulus of Neat epoxy & Composite samples treated at different conditions (Ductile Samples)

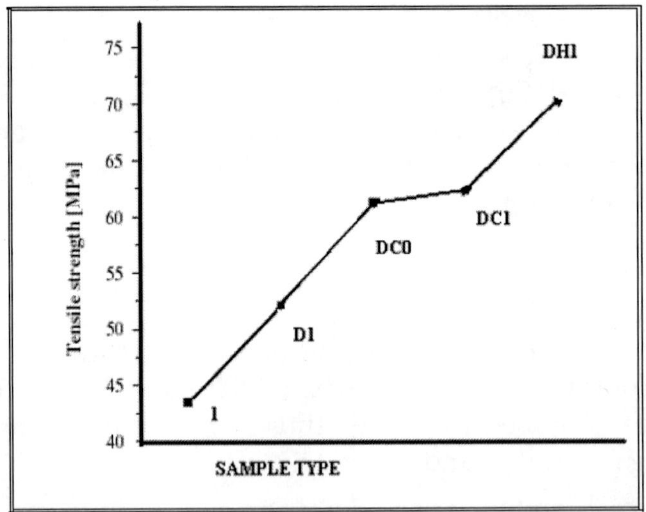

Figure 5. Tensile modulus of samples (1, D1, DC0, DC1, DH1)

The above mentioned relative improvement of epoxy resin in strength by adding CNTs can be explained by the high specific mechanical property and specific surface area of the MWCNTs [14]. Here, MWCNTs play the role of an enhancing framework under the treatment of hot water. The Young's modulus reduced simultaneously with thermal conditioning of the samples followed the reverse ordered followed by flexural modulus and Elastic modulus (Fig.6).

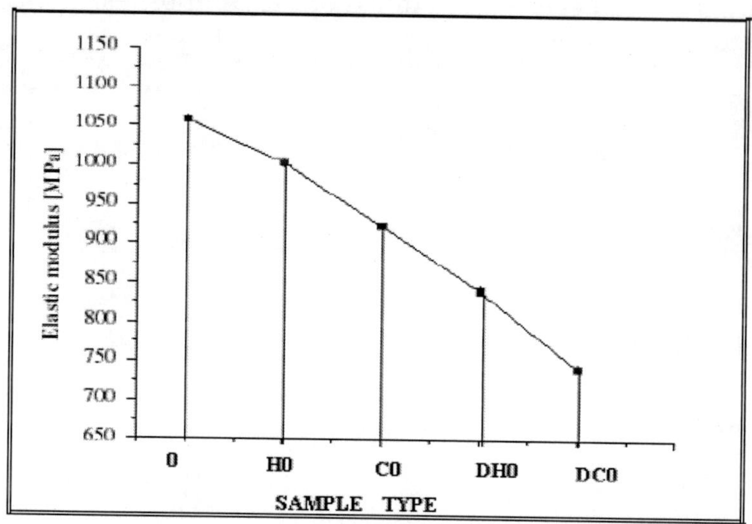

Figure 6. Elastic modulus of different samples (0, H0, C0, DH0 and DC0)

An explanation for the behavior can be found by considering the contraction and expansion of the matrix under high temperature and low temperature.

The flexural modulus was found to be 136.86 MPa in case of epoxy/RCNT composite which is about six times than the flexural modulus of plane epoxy sample (24.52 MPa). Increase in modulus is more pronounced in epoxy/ACNT composite i. e. 837.42 MPa which is more than six times that of epoxy/RCNT composite and thirty four times that of epoxy sample. This may be due to efficient load transfer from matrix to aligned CNT in Axial direction. Local stiffening due to nanotubes results in improved load transfer at the fibre/matrix interface. It had been reported that the increase in elastic modulus between the random and aligned nanocomposite is a consequence of the nanotube orientation, not polymer chain orientation. A considerable enhancement of hardness is observed by the nanocomposites in comparison to pure resin sample (Fig.7 b). Pure resin samples showed hardness of 12 MPa. Epoxy/RCNT had hardness value of 18 MPa which is 50% more and epoxy/ACNT had 49 MPa that is about four times that of epoxy sample. High strength and long nanotube reinforcements may result in forming a network structure that improves the hardness of the composites.

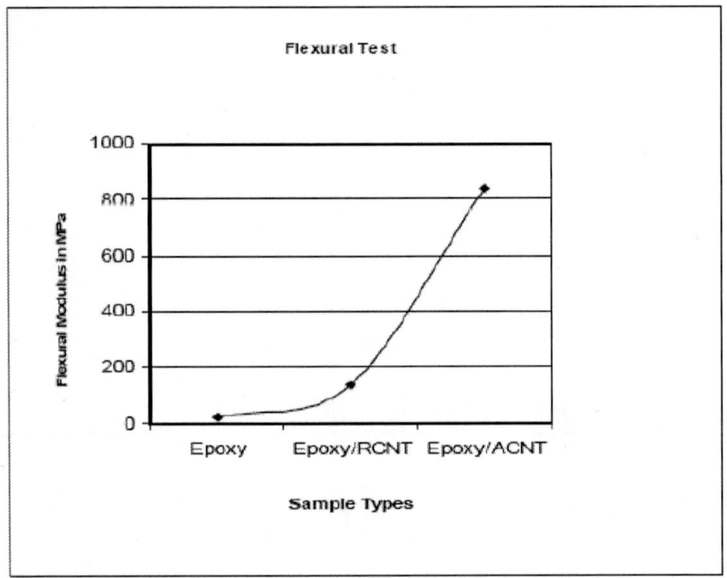

Figure 7. a Flexural Modulus profile

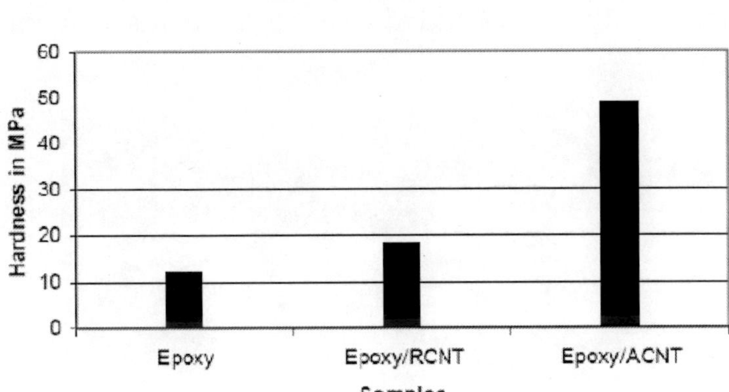

Figure 7. b Hardness profile

Fracture surface

Neat epoxy resin (Fig.8a) exhibits a relatively smooth fracture surface and the higher magnification SEM picture (Fig.8b) indicates a typical fractography feature of brittle fracture behavior, thus accounting for the low fracture toughness of the unfilled epoxy.

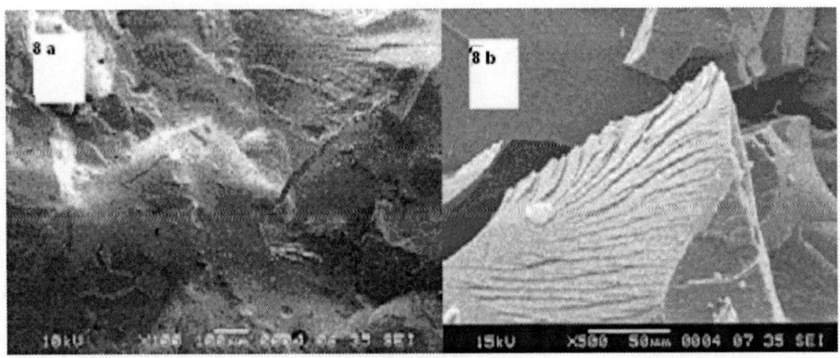

Figure 8 (a&b): Fracture surface of Epoxy resin at different magnifications

The distance between two cleavage steps (Fig.9a) is about 23-32 μm and the cleavage plane between them is flat and featureless. The fracture surfaces of the nanocomposites show considerably different fractographic features. The failure surfaces of the nanocomposites are rougher with the CNTs added into the epoxy matrix (Fig.9b).

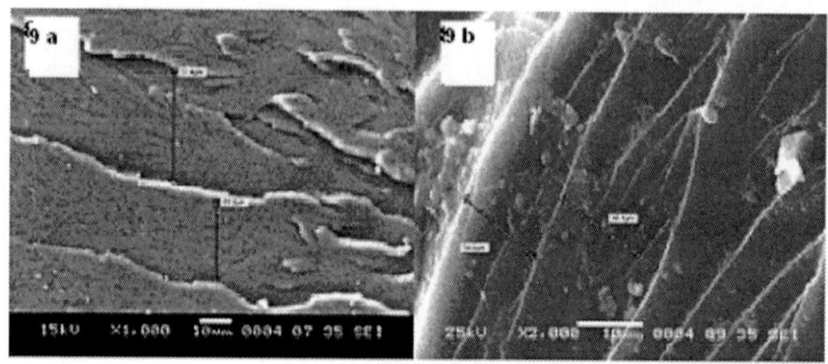

Figure 9. Cleavage plane of (a) pure epoxy (b) composite sample

The higher magnification SEM picture shows that the size of the cleavage plane decreased to 14-18 μm after the infusion of the CNTs. The decreased cleavage plane and the increased surface roughness imply that the path of the crack tip is distorted because of the carbon nanotubes, making crack propagation more difficult. Figure 9 shows the SEM images of the fracture surfaces of CNT/epoxy composite. The fracture surface of the CNT composite is very rough, which indicates that its failure was the result of a ductile deformation (Fig. 10a). However, the smooth fracture surface of the CNT composite suggests that it had a brittle failure mode (Fig. 10 b).

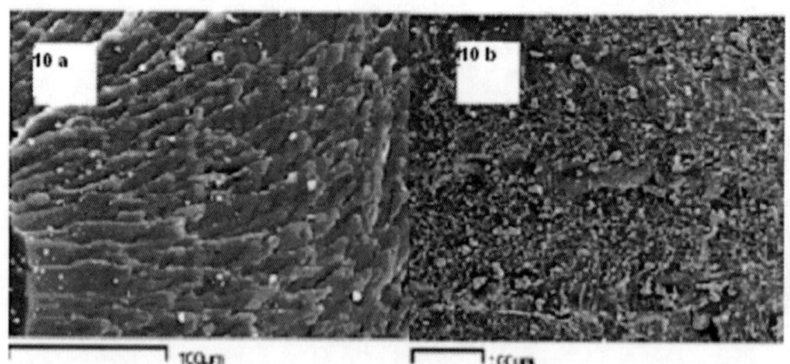

Figure 10. SEM images of the fracture surface of carbon nanotubes/epoxy composites. (a) Composites show a ductile deformation & (b) Composite with a brittle crack surface.

A large particle, an agglomeration of several carbon nanotubes (Fig.11 a, b, c) was observed in the fracture surface. But in Fig.11 (d), the agglomeration is less which results in the better mechanical properties.

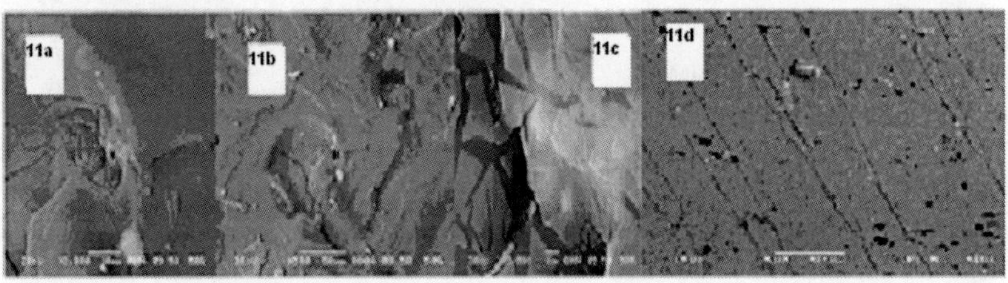

Figure 11(a, b, c & d). Agglomeration of several carbon nanotubes shown by black area in the centre

At a low stress level, the agglomerated particle increased the stiffness of the material, but at a high stress level, the stress concentration caused by the agglomerated particle initiated a crack, which made the sample fail quickly. Figure.12 (a) shows original traces of nanotubes in the composites. The fracture process did not follow the nanotube pullout pattern as in Fig. 12 (b), cracks propagated along the plane of the nanotube mesh. Higher magnification showed a crack interacting with the nanotube reinforcement. RCNT matrix pullout was observed along with extension and bridging of RCNTs across the crack. In epoxy/ACNT composites, the cracks were spanned by the nanotubes causing enhanced resistance to the crack propagation process. The bridging of the nanotubes as a mechanism of inhibiting the crack initiation in polymer and ceramic based nanocomposites has been well illustrated in literature[20-24].

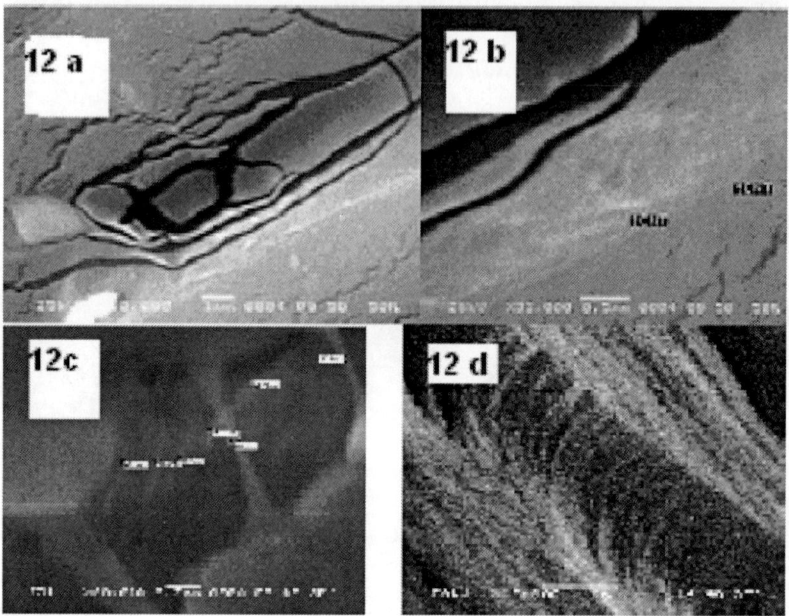

Figure 12 (a) Traces of nanotubes in the composite near crack, (b) crack propagation along the nanotube mesh, (c) Traces of stress concentrators in the composite near crack, (d) Show bridging by aligned nanotubes in the composites during deformation.

This agrees with the previous result submitted by the researchers that the all nanotubes, aligned close to the load direction in nanotube polymer composites were always pulled out at failure. This was a result of a poor interfacial bonding between the nanotubes and matrix. Therefore, the nanotubes inside the composites could not fully take up the load on the nanotube's longitudinal direction, which resulted in the decrease of flexural strength of the nanotube composite beams.

CONCLUSION

Addition of nanotubes enhanced the flexural & tensile properties because all the composite samples are showing better result than pure resin samples. Flexural modulus is maximum in case of ductile composite samples treated in hot water (DH1) at 800 C. These nanocomposites appeared tough while sub-ambient ductile samples (cryofreezer treated) are showing brittleness. The fracture surfaces of nanotube/polymer composites after flexural tests show different failure mechanisms for composites pre-treated under different conditions. The fracture process of composite beam appeared to indicate that failure was the result of agglomeration due to which crack initiation occurs.

In addition, aligned nanotube composites resulted in significantly improved flexural modulus and hardness indicating that there is efficient load transfer between the polymer matrix and the nanotube reinforcement along axial direction. Reduction in flexural modulus and hardness value in epoxy/ RCNT composites was due to formation of agglomerates of nanotubes inside polymer matrix that reduced reinforcing effects of the CNTs by acting as flaws in the resin. Investigation of fracture surface in nanocomposite revealed that narrower crack-tips underneath the advancing cracks were more efficiently bridged by the nanotubes in epoxy/ACNT resulting in an increased resistance against crack propagation.

REFERENCES

1. L. Cai, H. Tabata, and T. Kawai, 2000, "Self-assembled DNA networks and their electrical conductivity", *Appl. Phys. Lett.*, Vol. 77, pp 3105.
2. M. C. Hersam, A. C. F. Hoole, S. J. O'Shea and M. E. Welland, 1998, "Potentiometry and repair of electrically stressed nanowires using atomic force microscopy" *Appl. Phys. Lett.* , Vol.72, pp. 915.
3. G. B. M. Fiege, A. Altes, R. Heiderhoff and L. J. Balk, 1999 "Quantitative thermal conductivity measurements with nanometre resolution" *J. Physics,* vol.D32, pp. 113.
4. S. Gomes, N. Trannoy and P. Grossel, 1999 "DC thermal microscopy: study of the thermal exchange between a probe and a sample" *Meas. Sci. Technology,* vol.10, pp. 805.
5. O. Lourie and H. D. Wagner, 1998, *Journal of Matt Res*earch, vol.13, pp. 2418.
6. E. W. Wong, P. E. Sheehan and C. M. Lieber, 1997, "Nanobeam mechanics: Elasticity, strength, and toughness of nanorods and nanotubes", *Science*, vol. 277, pp. 1971.
7. J. B. Donnet, *Compos. Sci. Technol.* 2003. Vol. 63, pp. 1085–1088.
8. R. Bagheri and R. A. Pearson, 2000, *Polymer*, vol.41,pp. 269–276.
9. T. Kawaguchi and R. A. Pearson, 2003, *Polymer*.vol.44, pp. 4239–4247.
10. H. Mahfuz, A. Adnan, V. K. Rangari, S. Jeelani and B. Z. Jang, 2004, *Compos. Part A:*

11. Sci. Manuf., vol. 35, pp. 519–527.
12. M. F. Evora and A. Shukla, 2003, Mater Sci. Engg.A, vol.361, pp. 358–366
13. R. Rodgers, H. Mahfuz, V. Rangari, N. Chisholm and S. Jeelani, 2005, Macromol.
14. Mater. Engg, vol. 290, pp. 423–429.
15. P. Farhana, Y.X. Zhou, V. Rangari and S. Jeelani, 2005 *Mater. Sci. Eng. A,* vol. 405 (1–2), pp. 246–253.
16. Y. H. Liao, M. T. Olivier, Z. Y. Liang, C. Zhang and B. Wang, 2004, *Mater. Sci. Eng.A,* vol.385, pp. 175–181.
17. C. G. Zhao, G. J. Hu, R. Justice, D. W. Schaefer, S. Zhang, M. S. Yang and C. C. Han, 2005, *Polymer,* vol. 46, pp. 5125–5132.
18. . Kim, T. W. Pechar, E. Marand, 2006, *Desalination,* vol.192, pp. 330–339.
19. H. Cai, F. Y. Yan and Q. J. Xue, 2004, *Mater. Sci. Eng.* vol. 364, pp 94–100.
20. T. Ogasawara, Y. Ishida, T. Ishikawa and R. Yokota, 2004, *Compos. Part A: Appl.Sci. Manuf, vol.35,* pp. 67–74.
21. F.H. Gojny, J. Nastalczyk, Z. Roslaniec and K. Schulte, 2003, *Chem. Phys. Lett.,* vol. 370, pp 820–824.
22. H. Koerner, W. D. Liu, M. Alexander, P. Mirau, H. Dowty and R. A. Vaia, 2005, *Polymer,* Vol.46, pp. 4405–4420.
23. H. C. Kuan, C. M. Ma, W. P. Chang, S. M. Yuen, H. H. Wu and T. M. Lee, 2005 *Compos. Sci. Technology,* vol. 65, pp. 1703–1710.
24. M. K. Seo and S. J. Park, 2004, *Chem. Phys. Lett.,* vol. 395, pp. 44–48.
25. C. S. Li, T. X. Liang, W. Z. Lu, C. H. Tang, X. Q. Hu, M. S. Cao and J. Liang, 2004, *Compos. Sci. Technology,* vol.64, pp 2089–2096.
26. M. K. Seo, J. R. Lee and S. J. Park, 2005 *Mater. Sci. Eng. A,* vol. 404, pp 79–84.
27. J. Baschnagul, K. Binder, 1999, *MRS symposium. Proc*eeding,Vol. 543, pp. 157-164.
28. F.R Sherliker, 1998, *polymer Journal,* vol.12, pp.88.

Chapter 13

CHARACTERIZATION OF NANOTUBE-REINFORCED POLYMER COMPOSITES

Wenjie Wang[1] and N. Sanjeeva Murthy[2]

[1] Ames Laboratory, Iowa State University, Ames, IA, U.S.A.

[2] New Jersey Center for Biomaterials, Rutgers – The State University of New Jersey, NJ, U.S.A.

INTRODUCTION

Polymers are more readily processible than metals and ceramics, but are not as strong or as stiff. But the mechanical attributes of the polymer can be significantly improved by reinforcing polymers with high Young's modulus fibers. The resulting polymer composites are an important class of light weight materials with excellent specific mechanical properties. Unlike micron size particles and short fibers used in conventional composites, carbon nanotubes (CNTs) allow polymers to be reinforced at the molecular level and thus impart significantly greater improvement in mechanical properties. This is because for the reinforcement to be effective, it is necessary that the fibers be sufficiently long and the interface between the fiber and the matrix be strong, features that are provided by CNTs. The advantages of polymer-nanotube composites (PNCs) were discussed by Calvert (Calvert, 1999). Besides the economic advantage brought about by mixing costly CNTs and inexpensive polymer matrices, it is possible that a synergy exists between the CNTs and polymer matrices so that it is possible to go beyond the simple rule of mixture, and make full use of CNTs' exceptional mechanical properties to enhance the mechanical properties of the composite.

There are several reviews of PNCs (e.g., Breuer & Sundararaj, 2004, Moniruzzaman & Winey, 2006, and Coleman et al., 2006). The present chapter is different from these excellent reviews in that it focuses on the structure and morphology of the reinforced structures, and the consequence of these structures to their performance. The general features of the PNCs will be further illustrated using the results from our work on the study of reinforcements in polyacrylonitrile (PAN) with multi- and single-wall carbon nanotubes (CNTs).

POLYMER NANOCOMPOSITES

CARBON NANOTUBES AND NANOFIBERS

Carbon nanotubes (CNTs), discovered by Iijima in 1991, are seamless cylinders made of rolled up hexagonal network of carbon atoms (Iijima, 1991). Single-wall nanotubes (SWNTs) consist of a single cylindrical layer of carbon atoms. The details of the structure of multi-wall carbon nanotubes (MWNTs) are still being resolved, but can be envisioned as a tubular structure consisting of multiple walls with an inter-layer separation of 0.34 nm. The diameter for inner most tube is on the scale of nanometer, and the length of the tube can be up to ~ μm. In one model, the MWNTs are considered as a single sheet of graphene that is rolled up like a scroll to form multiple walls (Dravid et al., 1993). In another model, MWNT is thought to be made of co-axial walls consisting of hexagons aligned in helical manner, the individual tubes forming a nested-shell structure. Ge and Sattler (Ge & Sattler, 1993) found that as the graphene sheet rolls up and matches at joint line seamlessly, different wrapping angles resulting in different helicities is necessary for tubes to be separated by a uniform spacing. It is also likely that there could be the coexistence of scroll-type (open cylinder) and perfect Russian doll (nested shell) structures (Zhou et al., 1994). These two different structure models have an impact on the mechanical as well as the optical and electronic properties of CNTs. The carbon atoms in the plane of graphene sheet are bound with strong covalent sp^2 hybridized bonds. The interaction between walls can be approximated by weak van de Waals force or π bond. Thus, the sliding of walls with respect to each other might significantly reduce axial modulus of MWNT compared to SWNT, more so in the nested shell structure than in the scroll-type structure.

Vapor-grown carbon nano-fibers (VGCNF), which are also used as reinforcements, are 50-200 nm in diameter, 20-100 μm in length, and bear resemblance to MWNT, except that graphite sheets make an angle (~15°) to the long- axis (Uchida et al., 2006). Compared to the CNTs, VGCNF is less costly to produce.

CNTs have exceptional mechanical, optical, electrical and thermal properties (Chae et al., 2009; Chae et al., 2007; Guo et al., 2005; Koganemaru et al., 2004; Sreekumar et al., 2004; Ye et al., 2004). The elastic modulus of an isolated CNT, estimated using intrinsic thermal vibrations in TEM, is 1.8 TPa. (Treacy et al., 1996), almost an order of magnitude larger than that of carbon fibers. This axial modulus is higher than the 1 Tpa in-plane modulus of a single graphene sheet due to the wrap-up of the graphene sheet into a cylinder. Bower et al. measured the fracture strain of MWNT ≥18% in polymer composites, almost one or two orders of magnitude larger than carbon fibers (Bower et al., 1999). Chae and Kumar suggested that, because the tensile strength of CNTs is 100-600 GPa, about two orders of magnitude larger than that of ordinary carbon fibers, the CNTs might be the ultimate reinforcing materials for strong fibers (Chae & Kumar, 2009).

CNT/POLYMER COMPOSITES

The advantages of using CNTs as reinforcements in polymer matrices can be illustrated with the following empirical relationship between the critical length (lc), and fracture stress σf of the reinforcing fiber

(Wang et al., 2008a): *lc=σfd2τc*

where *d* is the fiber diameter, and τ_c is the interfacial strength. The critical fiber length is the minimum length of the fiber such that the failure occurs within the fiber, and thus ensures that it acts as an effective reinforcement. This relationship shows that nanotubes with their smaller diameter, and greater interfacial strength when molecularly dispersed serve as effective reinforcements even at smaller lengths.

The importance of bonding between CNTs and matrices for effective load transfer without slipping of contact surfaces has been widely recognized (Calvert, 1999). It is desirable that a large fraction of the load applied to the composites is efficiently transferred to CNTs through the interaction or bonding between the outer surface of the CNTs and the surrounding polymer matrix (Ajayan et al., 1997). It is widely accepted that larger the area of interface between the matrix and nano-fillers, the better the mechanical enhancement. Several load-transfer models have been proposed and examined using atomistic molecular dynamic simulations. Most recent molecular dynamics simulations suggest that weak van der Waals force (Srivastava et al., 2003), as well as thermal expansion mismatch can contribute to strong interfacial interaction (Wong et al., 2003; Zhang & Wang, 2005). Panhuis's model suggests that the polymer chains will adapt their conformations to wrap around a single CNT (Panhuis et al., 2003; Wong et al., 2003). Frankland's model suggests that a small fraction of chemical bonds are formed between polymers and carbon atoms of the CNTs thus cross-links are formed between the matrix and CNTs (Frankland et al., 2002).

Chemical bonding between CNT and polymer is apparent in the adhesion of polymers to CNT on the fracture surface through transmission electron microscopy (TEM) (Bower et al., 1999). However, Raman spectroscopy along with the SEM images of fracture surface by Ajayan et al. showed the sliding of SWNT bundles with respect to the matrix suggesting that the actual load transfer was poor between CNT and the polymer matrix, which rules out the chemical bonding model (Ajayan et al., 2000). A number of experiments have been carried out to test these models for the interaction between polymer matrix and CNTs (SWNTs and MWNTs) through tensile tests, microscopy and x-ray scattering. Deformation of the CNTs observed by x-ray scattering measurements (see section 3.3) shows an interaction between the matrix and CNTs that is sufficient to transfer the load from the matrix to the CNTs (Wang et al.,2008b). Actual measurements of the adhesion between the matrix and CNTs were also carried out using scanning probe microscope to pull out the CNTs from the matrix (Cooper et al., 2002). Coleman et al. found that a layer of polymer attached to the CNTs after pullout was very similar to the crystalline polymer coating nucleated by the MWNTs. The presence of interfacial crystalline layers partly contributes to the mechanical properties of composites (Coleman et al., 2006). Pull-out measurement can also be carried out by atomic force microscopy (AFM). Barber et al. attached one end of a single CNT to the tip of AFM and embedded the other end into the polymer matrix. The critical force needed to pull the nanotube from the polymer matrix was then recorded (Figure 1).

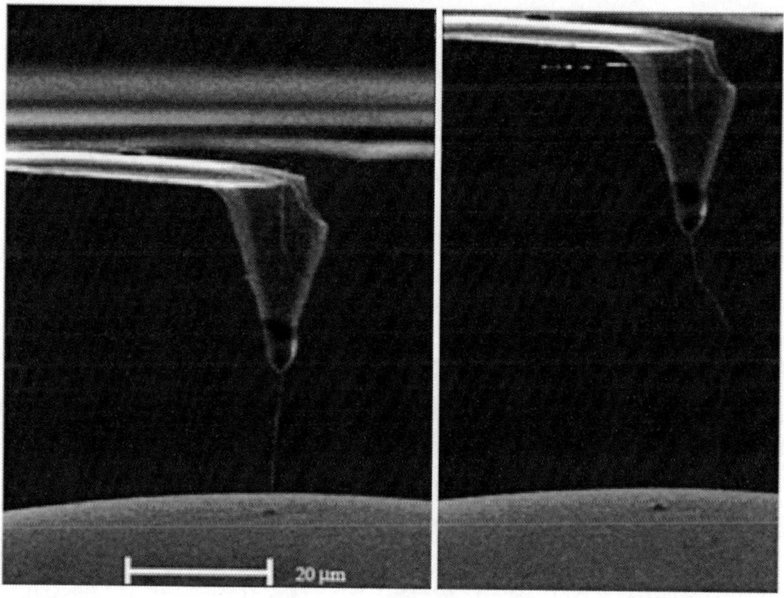

Figure 1.

Left panel shows that one end of a single CNT is attached to the AFM tip and the other is embedded in the solid polymer matrix.

Right panel shows the process of pull-out of CNT. The critical force needed for pull-out can be recorded by AFM (Reprinted from Philosophical Transactions of the Royal Society A: Mathematical, Physical and Engineering Sciences, vol. 366, W. Wang, P. Ciselli, E. Kuznetsov, T. Peijs and A.H. Barber. "Effective reinforcement in carbon nanotube-polymer composites", Figure 2, Copyright 2008, with permission from the Royal Society.).

Because of the interfacial interaction between CNTs and the polymer matrix, the properties of the matrix in the interfacial regions are different from the bulk materials (Schadler, 2007). CNTs dispersed in the polymer matrix, despite the small volume fraction (usually 1-5 vol%), enhance the overall mechanical properties, also affect the structure of the matrix during conventional processing. Many observations show that the presence of CNTs can affect the crystallization behaviors of the polymers (Brosse et al., 2008; Grady et al., 2002; Liu et al., 2004) and alter the morphology of matrix (Brosse et al., 2008; Chatterjee et al., 2007). For instance, instead of the expected spherulites, crystalline lamellae were observed growing perpendicular to the surface of CNTs. (Brosse et al., 2008). Therefore, the microstructure of the polymer matrix should be taken into account when evaluating the performance of nanocomposites (Advani, 2006).

Two other factors that affect the performance of CNT-reinforced nanocomposites are the degree of dispersion and alignment of nanotubes. Poor or nonuniform dispersion causes the fillers to aggregate, acting as defect sites (Advani, 2006). Dispersion process is affected by interactions of CNT-CNT and CNT-matrix. It is controlled by several factors such as filler's size, specific surface area, and interfacial volumes (Moniruzzaman & Winey, 2006). The degree of dispersion of CNTs in polymer matrix can be assessed using small angle x-ray scattering as described in the following section. Rheological study of dispersions of MWNTs in the polymer matrix by Huang et al (Huang et al., 2006) also suggests that a satisfactory dispersion can be achieved if the mixing time exceeds a critical value (characteristic mixing time).

PROCESSING OF PNCS

CNTs are dispersed into the polymer matrix to form a polymer/CNTs composite, and there are several ways to achieve a good dispersion.

Following is a condensed version of the various processing methods as discussed by Coleman et al. (Coleman et al., 2006). Solution processing is the easiest if the polymer can be dissolved in a suitable solvent. Typically, nanotubes well dispersed in a solvent is mixed with a polymer solution with energetic agitation (e.g., sonication), and the solvent is allowed to evaporate in a controlled way leaving a composite film. For insoluble thermoplastic polymers, the polymer is heated above either glass transition (if amorphous) or melt temperature (if semicrystalline) to form a viscous liquid, and the CNTs are mixed into this viscous liquid to be then processed into a fiber or a film using common production processes. For thermosetting polymers such as epoxy, the CNTs are blended when epoxy is in a liquid state, and the mixture is then allowed to cure. Chemical processing of Polymer/CNTs composites aims at formation of chemical bonds between the polymer matrix and the CNTs; this provides for stronger interaction between the polymer and CNTs and better dispersion, and thus results in better mechanical properties. Chemical methods include *in-situ*polymerization and functionalization of CNTs with polymer, also called polymer grafting.

Characterization of CNT/Polymer composites

Although the CNTs' high strength and stiffness as well as high aspect ratio promise significant improvement in mechanical properties of polymer nanocomposites (Treacy et al., 1996), the actual measured enhancement is below these expectations (Schaefer et al., 2003). The major reason might be the intrinsic van der Waals' attraction makes CNTs entangled agglomerates incapable of fully blending into the polymer matrix. Since the four factors important to the reinforcement are CNTs' aspect ratio, alignment, dispersion, and interfacial interactions (Coleman et al., 2006), it is desirable to examine the morphology of CNTs in the polymer matrix. Several methods including x-ray diffraction, Raman spectroscopy, scanning and transmission electron microscopy are employed to characterize the structure, morphology and properties of carbon nanotube-reinforced polymer nanocomposites.

ELECTRON MICROSCOPY

Transmission electron microscopy (TEM) and scanning electron microscopy (SEM) enable the visualization of the microscopic structure and morphology of CNTs in composites at resolution down to 0.1 nm (Michler, 2008; Sawyer et al., 2010). The limitations of using electron beam to probe the polymer are the low contrast (most polymers are made of light elements) and radiation damage. TEM is used to probe the bulk structure from thin sections of the material, and SEM to examine the surfaces. In addition to imaging, TEM is also used to carry out micromechanical measurements. Examples of these applications will be illustrated below.

DISPERSION AND ALIGNMENT OF THE CNTS

One of the primary uses of microscopy is characterizing the dispersion and alignment of the nanotubes in the polymers. Figure 2a is an SEM image from a composite prepared by mixing aqueous solution poly(vinyl alcohol) with CNT dispersions followed by subsequent casting and controlled water evaporation (Shaffer & Windle, 1999). It shows homogeneous dispersion of the nanotubes even at 50 wt% loading. By examining the fracture surface using SEM, they confirmed the uniform dispersion of CNTs. Figure 2b shows the use of TEM to examine the dispersion of CNTs (Qian et al., 2000). In this experiment Qian et al. used high-energy sonication to disperse MWNTs (predispersed in toluene) into polystyrene (PS) dissolved in toluene, and the solvent was then evaporated to produce 0.4 mm thick films. Figure 2b shows 34 nm diameter MWNTs well dispersed down to the μm length scales. The inset is a plot of the loading of the CNT's at different length scale, and shows that homogeneity increases (decrease in standard deviation) with increase in the length scale. Furthermore, the TEM images also reveal defects that were present as a result of the polymer shrinkage primarily at the end of the nanotubes. Other than the quality of dispersion, both of these nanoscale images show that in PNCs, as pointed out by Fisher et al., the nanotubes are curved instead of being straight. Based on these observations, Fisher et al. illustrated that the reduction of effective modulus of CNT-reinforced materials is due to the curvature of embedded CNTs (Fisher et al., 2002).

Figure 3 illustrates the degree of alignment that can be achieved in PNC composites as seen in a TEM image (Thostenson & Chou, 2002). Unlike the two examples discussed above, in this instance, Thostenson & Chou extruded the polymer (polystyrene, PS) and a master batch of nanotube dispersed in PS using a micro-compounder. The film coming out of a rectangular die was drawn and solidified by passing over a chill roll. TEM was used to characterize the dispersion and alignment of MWNTs in polymer films. Figure 3a shows the large-scale dispersion and overall alignment of the carbon nanotubes and Figure 3b shows the alignment of the individual nanotubes. Examination of the drawn films showed that a draw ratio of five was sufficient to achieve a good nanotube alignment.

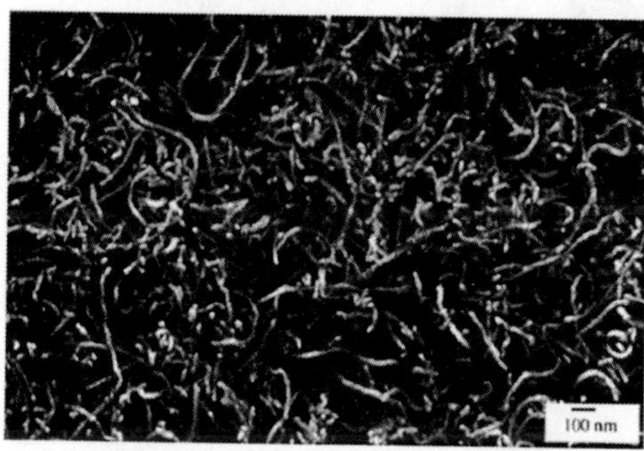

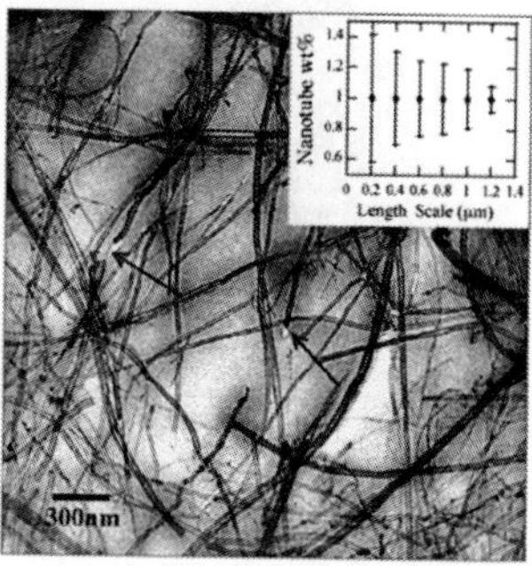

Figure 2.

Left Panel (a): A SEM image for 50 wt% CNT/ Poly(vinyl alcohol) composites (S.P. Shaffer and A.H. Windle, "Fabrication and characterization of carbon nanotube/poly(vinyl alcohol) composites". Advanced materials (1999). Vol. 11. 937-941. Copyright Wiley-VCH Verlag GmbH & Co. KGaA. Reproduced with permission.). Right panel (b): An example of TEM image at 1 μm length scale showing homogeneously dispersed MWNT in polystyrene (PS) matrix. Homogeneity of nanotube dispersion was evaluated by assessing the weight fraction of nanotubes in different areas in the TEM images for different length scale. The inset shows that homogeneity is greater at 1.2 μm length scale (small error bar) than lower length scale. An arrow in the image indicates the defects present at the nanotube ends

(Reprinted with permission from D.Qian, E.C. Dickey, R. Andrews and T. Rantell., "Load transfer and deformation mechanisms in carbon nanotube-polystyrene composites", Applied Physics Letters, vol. 76, 2868-2870. Copyright 2000, American Institute of Physics.)

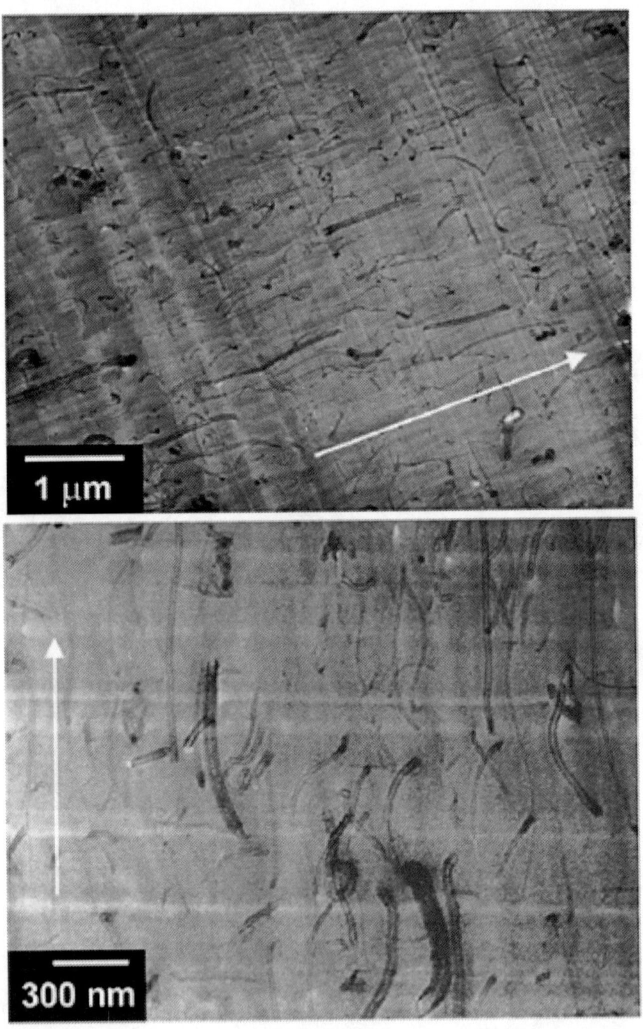

Figure 3.

TEM images of dispersion and alignment of CNTs in nanocomposites films. The arrows indicate flow/drawing direction. Left panel (a): TEM at length scale of micrometer. Right panel (b): TEM at length scale of nanometer (Reprinted with permission from

E.T. Thostenson and T.W. Chou, "Aligned multi-walled carbon nanotube-reinforced composites: processing and mechanical characterization", J. Phys. D: Appl. Phys., vol. 35, L77-L80. Copyright 2002, Institute of Physics Publishing.).

MECHANICAL PROPERTIES OF CNTS

TEM is also used to measure the mechanical properties and to study the deformation mechanism in PNCs. In fact, the Young's modulus of CNTs, which is hard to measure by conventional means, was first directly measured using TEM from the thermal vibration amplitudes of CNTs, and thus confirming the predicted axial stiffness on the scale of terapascal (TPa) and the high axial strength due to the in-plane graphitic structure of CNTs (Treacy et al., 1996). Bower et al. (Bower et al., 1999) aligned MWNTs in the thermoplastic polymer (polyhydroxyaminoether, DowChemical Co.) films by stretching above the glass transition temperature and releasing the load at 300 K. The uniaxial alignment of MWNTs was confirmed by x-ray diffraction. The CNTs in the polymer matrix are bent (Figure 4), presumably because of the shrinkage of the polymer matrix as the composite is cooled from 100° C to room temperature. The deformation was reversible at moderate bending, i.e. the CNTs returned to their straight cylindrical shape by heating the polymer matrix. Fracture surfaces in PNC's can be examined to ascertain the strength of the adhesion between the polymer and the nanotube. Figure 5 shows MWNT protruding from polymer matrix attached with a layer of polymer. All these observations point to the effective interactions and intimate contact between CNTs and polymer matrix. In a final example, Figure 6 shows the use of TEM to observe the nucleation and propagation of a crack induced in thin MWNT-PS film by thermal stresses in a TEM (Qian et al., 2000). The nucleation and propagation occur in regions of low CNT density and along the weak CNT-polymer interface. Also, the CNTs tend to align perpendicular to the crack and bridge the gap, and some of them are already pulled out, suggesting that the critical failure force is determined by interaction between polymer matrix and CNTs. These CNTs provide a closure force until a critical value is reached to be pulled out of the matrix.

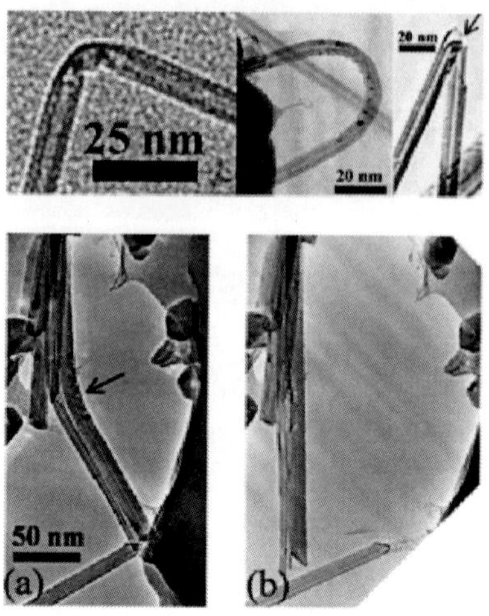

Figure 4.

Top panel: TEM images showing the buckling of in bent MWNTs in MWNT/polymer composites. Bottom panel: (a) A bent nanotube before heating; (b) the bent nanotube restored to its straight cylindrical shape after heating (Reprinted with permission from C. Bower, R. Rosen, L. Jin, J. Han and O. Zhou, "Deformation of carbon nanotubes in nanotube-polymer composites", Applied Physics Letters, vol. 74, 3317-3319. Copyright 1999, American Institute of Physics.).

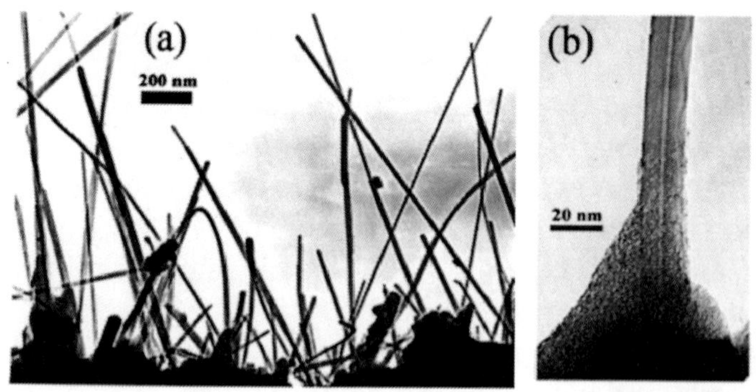

Figure 5.

TEM images of the fracture surface showing MWNT protruding from the polymer matrix suggestive of strong adhesion between the

MWNT and the polymer matrix (Reprinted with permission from C. Bower, R. Rosen, L. Jin, J. Han and O. Zhou, "Deformation of carbon nanotubes in nanotube-polymer composites", Applied Physics Letters, vol. 74, 3317-3319. Copyright 1999, American Institute of Physics.).

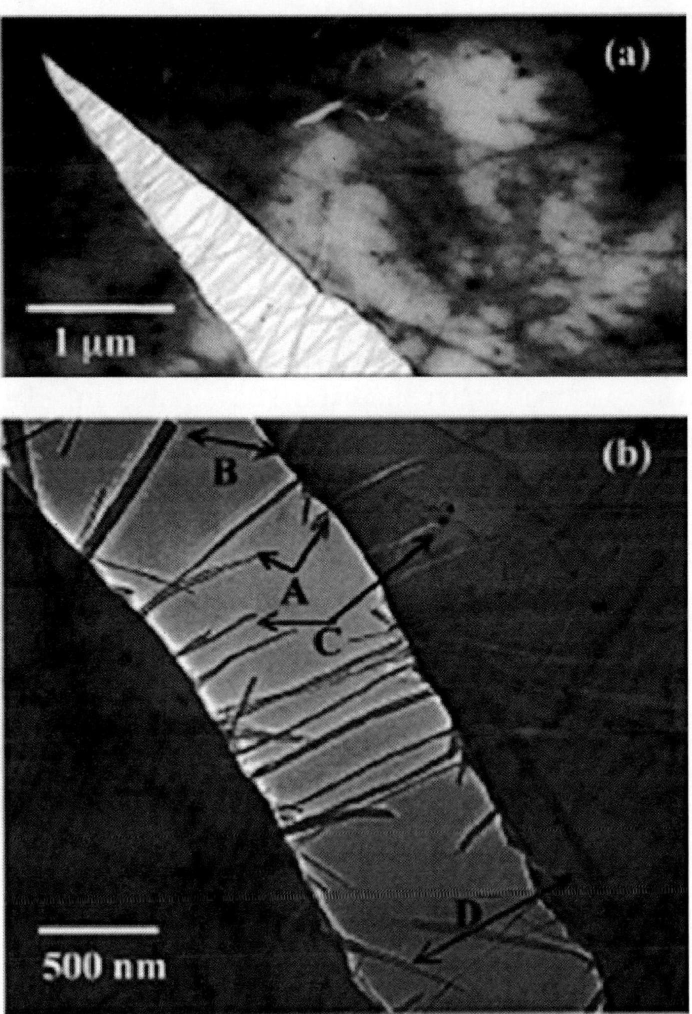

Figure 6.

Crack initiated by thermal stress. Upper panel shows the crack initiation and propagation at low CNT density areas. Lower panel shows the CNTs aligned perpendicular to the crack bridging the crack. Arrows labelled A, B, C and D point to CNTs that are broken and pulled out of matrix (Reprinted with permission from D. Qian, E.C. Dickey, R. Andrews and T. Rantell, "Load transfer and deformation mechanisms in carbon nanotube-polystyrene composites", Applied Physics Letters, vol. 76, 2868-2870. Copyright 2000, American Institute of Physics.).

RAMAN SPECTROSCOPY

Infrared (IR) and Raman spectroscopy (RS) are the two vibrational spectroscopic techniques widely used to obtain molecular structure information for the identification of polymers, functional groups and stereochemical structures. Both techniques measure the vibrational energies of molecules. However, for a vibrational motion to be IR active, the dipole moment of the molecule must change, and for a transition to be Raman active there must be a change in polarizability of the molecule. In most instances, IR and RS yield complementary information because IR activity suggests RS inactivity and vice versa. Sample requirements for RS are less stringent than for IR. Details of application of RS to CNTs have been discussed thoroughly in the literature (Dresselhaus et al., 2005), and a few examples of the use of RS to PNCs are provided here. For oriented specimen, the intensity of RS spectra line depends on the orientation of molecules relative to the polarization of the incident beam as well as scattering beam. This feature is used to determine the orientation distribution for CNTs in polymer matrix. Polarized RS has also been used to probe the one -dimensional nature of SWNT in the bulk materials (Gommans et al., 2000). These results show that the RS intensity has the same orientation dependence as nanotube's polarizability. A similar work on MWNT has also shown that there is a strong dependence of the graphite-like G band and the disorder-induced D band on the polarization geometry (Rao et al., 2000).

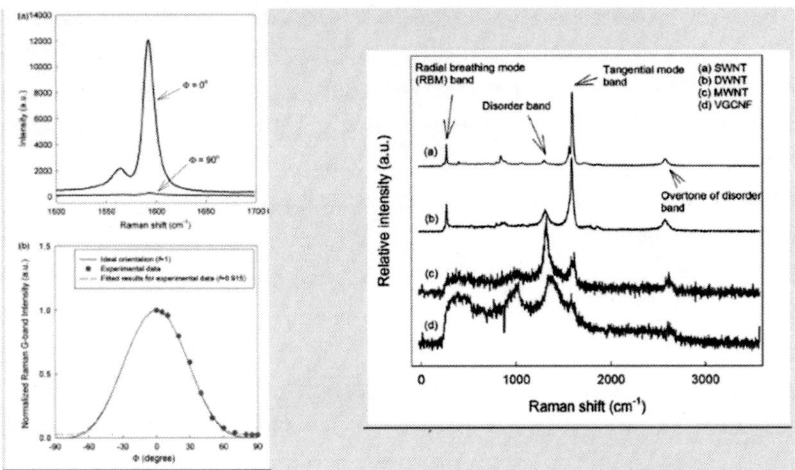

Figure 7.

Upper left panel: G-band Raman spectra for PAN/SWNT fiber. The ratio of the peak intensity at 1592 cm^{-1} are compared when the angles between polarizer and the fiber axis are 0 and 90°, respectively. Lower left panel: The normalized G-band Raman intensity as function of angles between polarizer and fiber axis. Experimental data are for PAN/SWNT (1 wt%) at draw ratio 51 (Reprinted from Polymer, Vol.47, H.G. Chae, M.L. Minus and S. Kumar, "Oriented and exfoliated single wall carbon nanotubes in polyacrylonitrile", 3494-4504. Copyright 2006, with permission from Elsevier.) Right panel: Raman spectra for pristine CNT powders (Reprinted from Polymer, Vol.46, H.G. Chae, T.V. Sreekumar, T. Uchida and S. Kumar, "A comparison of reinforcement efficiency of various types of carbon nanotubes in polyacrylonitrile fiber", 10925-10935. Copyright 2005, with permission from Elsevier.).

Mathematical description of Raman intensity distribution as a function of SWNT's orientation has been derived using Legendre polynomials by Liu et al. (Liu & Kumar, 2003). They built a relationship between 2nd and 4th orientation parameters of SWNT and Raman scattering intensity and depolarization ratio. This quantitative relation was used by Chae et al. to investigate the orientation of SWNT in polyacrylonitrile fibers (Chae et al., 2006). They measured G-band intensity ratio at about 1592 cm^{-1} for polarization parallel and perpendicular to the fiber axis (Figure 7). By comparing the G-band ratio for conventional spun and gel-spun PAN/SWNT fibers, they were able to conclude that the orientation of SWNT is only slightly improved by the gel-spinning process. They also showed that RS could be used to estimate the relative intensity of disorder band for SWNT, DWNT, MWNT, VGCNF (Chae et al., 2005) (Figure 7). They concluded that SWNT has best perfection while MWNT and VGCNF have more disorder (Chae et al., 2005).

X-RAY SCATTERING

X-ray scattering is an ideal tool to investigate the influence of CNTs on polymer morphology at multiple length scales from nm to sub μm length scales. This is achieved by using variations of the basic diffraction/scattering technique that are usually referred to as x-ray diffraction (XRD), wide-angle x-ray diffraction (WAXD), and small-angle x-ray scattering (SAXS). The scattering or diffraction intensity is expressed as a function of the scattering vector, $q(=4\pi\sin\theta/\lambda)$ or $s(=2\sin\theta/\lambda)$, 2θ being the scattering angle, and λ the x-ray wavelength. The sample preparation is far simpler than that for TEM. Whereas electron microscopy can provide information about the distribution of the CNTs and other structural information within a particular frame of an image, SAXS can provide similar data about the distribution, but averaged over lengths ~ mm in the bulk of the sample. Some of the characteristics of the PNCs that can be measured using x-ray scattering include the structural changes in the polymer due to the incorporation of the CNT, orientation of the polymer chains/crystals and the CNTs, and structure at the interface between the polymer and the CNT, and the changes in these features during deformation.

The advent of synchrotron radiation, whose brilliance is ~ 10 orders of magnitude higher than that from a rotating anode tube, makes it possible to carry out *in-situ* magnitude higher than that from a rotating anode tube, makes it possible to carry out *in-situ* dynamical measurements. Some of the data in this section were obtained from CNTs dispersed in polyacrylonitrile (PAN) that was then spun into a fiber. PAN fibers reinforced with CNTs are made by solution- or gel-spinning. PAN/CNTs fibers with significant enhancement in properties suggested good interactions between PAN and CNTs

(Chae et al., 2006; Chae et al., 2005; Guo et al., 2005; Sreekumar et al., 2004; Uchida et al., 2006). These studies showed that the improvement in low strain is due to the interaction of PAN and CNTs, while the improvement as high strain is partly due to the CNT length.

CHARACTERIZATION OF THE CNTS

Before discussing the results for the PNCs, we will first discuss some of the scattering data relevant to the characterization for the CNTs. Historically, the MWNTs were observed by electron microscopy and were depicted as concentric seamless cylinders (Iijima, 1991). This model was used to interpret XRD data (Pasqualini, 1997), and further direct evidence appeared later to support this concentric cylinder model (Xu et al., 2001). However, there has been evidence that suggest that the local structure is similar to turbostratic graphite or scrolls of graphite sheets (Dravid et al., 1993; Zhou et al., 1994). More recent data suggest that while thin MWNTs could be modelled as concentric 7-15 tubules, thicker MWNTs were found to be mixtures of both scroll-type MWNT and concentric-cylinder (nested Russian doll) MWNTs (Maniwa et al., 2001).

An individual CNT cannot be characterized using SAXS. However, due to the π-π chemical bonding and van de Waals interactions, the CNTs tend to aggregate into bundles while lacking a clear-cut rod-like morphology (Schaefer et al., 2003a), and this typical morphology can be characterized using SAXS. Schaefer et al. combined small-angle light scattering, ultra-SAXS and SAXS to measure morphology and degree of dispersion of SWNTs and MWNTs suspended in water (Schaefer, 2003a;Schaefer, 2003b). They found that the intensities over 5 decades of q are characterized by different power laws. The power-law of intensity from SWNT indicates a branched rope network with fractal characteristics. Porod's law was observed at length scale smaller than inner radius, suggesting the smoothness of interface. The nonuniform dispersion of CNTs in polymers partly explains why the reinforcement of CNTs to polymer composites falls below expectations.

In another experiment (Inada et al., 2005), two segments of different power-law behavior were observed in the intensity distribution

(Figure 8). One segment conforms to Porod's law corresponding to smooth interfaces. The other follows a power-law with exponent -1, suggesting scattering is from a rod-like structure. The scattering could be modeled as those from MWNTs represented as hollow tubes with an inner diameter of about 8 nm and a wall thickness of 6 nm, which is in agreement with the reported results from direct measurement through TEM.

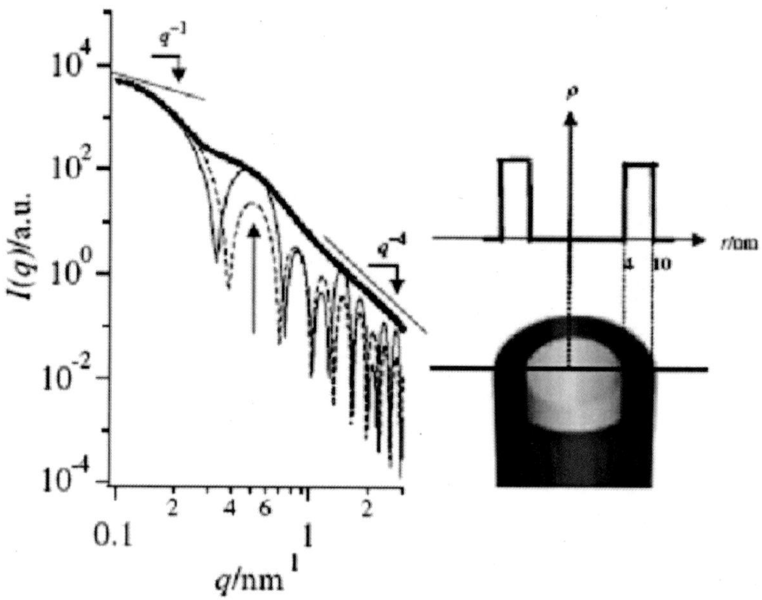

Figure 8.

Left: Measured SAXS intensity as function of scattering vector. Two power laws were presented indicating a one-dimension cylinder structure and sharp interface, respectively. Straight line and dotted line are calculated for a core-shell and hard cylinder models, respectively. Right: The best fit parameter suggest the structure of CNT is a core-shell structure (Reprinted from T. Inada et al., "Small-angle x-ray scattering from multi-wall carbon nanotubes (CNTs) dispersed in polymeric matrix", Chemistry Letters, Vol. 34, 525 (2005), with permission from The Chemical Society of Japan.)

SAXS CHARACTERIZATION OF THE VOIDS AND THE CNTS IN THE POLYMER

Voids are invariably formed in the fibers during the spinning and the drawing process, especially in the presence of reinforcements. Voids, and in general any scattering entities, give rise to SAXS that can be analyzed to obtain information about the length, diameter, and orientation of these entities as well the information about the nature of the interface. In the work with PAN fibers, the degree of orientation of the voids measured by SAXS was found to be same as those of the crystalline domains observed in WAXS. Both increased reversibly upon stretching, indicating that the voids are integral parts of the polymer matrix and closely associated with the crystalline domains in the fibrils (Wang et al., 2009). SAXS data also showed large changes in the size distribution of the voids during elongation in the solution spun PAN fibers, but not the composite fibers. Furthermore, heating the fiber above the glass transition temperature (*Tg*) decreased the volume fraction of the voids considerably as indicated by a decrease in the scattered intensity. Figure 9 shows examples of the decay in equatorial scattering with scattering angle. The rate of this decay, as indicated by the slope of the log-log plot (called the exponent) provides valuable insight into the nature of the void-polymer interface. In the plot of the PAN with MWNT, we see the exponent is 4.4 (Figure 9a). An exponent of 4.0 would indicate a sharp, smooth interface.

A value higher than 4 suggests that in the PAN/MWNT sample, there is a diffuse interface between the polymer and the void as indicated schematically in the figure. An exponent between 3 and 4 would indicate a rough void-polymer interface, which was observed in gel-spun fibers without any CNT's (Figure 9b). An exponent between 2 and 3 indicates that the structure within the void is similar to a mass-fractal. It is speculated that because of the nature of gel-spinning process, the voids are partially filled with polymer, and perhaps CNT, and as a result the scattering exponent falls below 3 as indicated in Figure 9c. It is possible that a similar analysis can be carried out to probe the nanotube-polymer interface provided there is a contrast between the CNTs and the matrix.

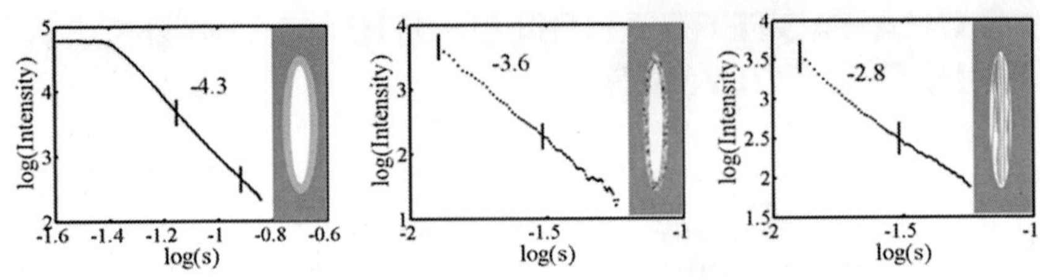

Figure 9.

Power law for equatorial intensities. (a) PAN/MWNT for a void with diffuse boundary ; (b) gel-spun PAN for a void with surface fractal; (c) gel-spun PAN/SWNT for a void with mass fractal The linear segments that are fitted to obtain the fractal dimension are shown in the figure (Wang et al., 2009).

WAXD ANALYSIS OF THE POLYMER AND THE CNTS

Figure 10 shows example of the 2D diffraction patterns and the corresponding 1D scans that can be obtained from polymer composites with CNTs. Because of the large fraction of the polymer (95%) the major features in these scans are due to the polymer, and these features can be used to follow the changes in the polymer structure in terms of the unit cell dimensions, crystallinity, crystallite size and polymer orientation (Wang et al., 2008b). For example, the data show that a monotonic decrease in PAN inter-chain spacing with the fiber strain is accompanied by a reversible change from helix to zigzag conformation, and by changes in the axial repeats of the two conformations. These changes are much larger in gel-spun fibers than in solution-spun fibers, indicating more effective load transfer between the amorphous and crystalline segments in gel -spun fibers. In addition to the intense polymer peaks, it is also possible to see the weak features of the CNT reinforcements in the pattern, especially in the 1D scans. Note the broad feature due to MWNT in Figure 10e, and the sharp features due to VGCNF in Figure 10f. The changes in these features with stress were used to ascertain the modulus of the polymer and the CNT in the composite. Such data are potentially useful in determining the distribution of load between the two components.

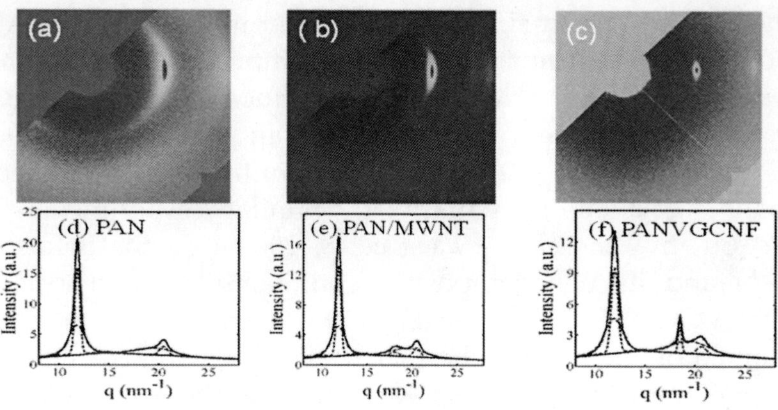

Figure 10.

Upper Panel: Two dimensional wide-angle x-ray diffraction pattern for (a) solution-spun PAN ; (b) solution-spun PAN/MWNT; (c) gel-spun PAN/SWNT fibers. Lower Panel: Profile fitted equatorial scans: (d) PAN; (e) PAN/MWNT; (f) PAN/VGCNF. The data for gel fibers are similar to that of PAN. (Reprinted from Polymer, Vol.49, W. Wang, N.S. Murthy, H.G. Chae and S. Kumar, "Structural changes during deformation in carbon nanotube-reinforced polyacrylonitrile fibers", 2133-2145. Copyright 2008, with permission from Elsevier.)

ALIGNMENT OF THE CNTS AND THE POLYMER

As indicated earlier, to a large extent, the strength, modulus and shrinkage properties of the polymer composite are determined by the distribution in orientation and alignment of both the CNT and the polymer. XRD is an appropriate tool for determining this parameter. (Wang et al., 2008b). For instance, Pichot et al. obtained the orientation of SWNT and PVA chain by azimuthal intensity distribution (Pichot et al., 2006). They related the alignment of SWNT in polyvinyl alcohol (PVA) fibers with different draw ratio and accounted for the improvement of alignment with draw ratio by applying a simplistic continuum mechanical model, i.e. the induced alignment is due to the constant volume. In the work on PAN fibers, by measuring the orientation during loading and unloading of the fibers, it was found that the orientation reversibly increases as the fiber is stressed (Wang et al., 2008b).

Chen et al. (Chen et al., 2006) employed in-situ WAXS/SAXS to investigate the deformation of nanocomposites fiber containing fluorinated MWNT (FMWNT) and fluorinated polyethylene-propylene. Orientation of FMWNT can be characterized by (002) reflection of WAXS and quantitatively described by Hermans' orientation parameter. Similar experiments were carried out byWang et al. (Wang et al., 2008b) on the deformation of PAN and its nanocomposites containing a variety of CNTs (MWNT,VGCNF,SWNT). The effect of CNT reinforcement could be ascertained by monitoring the structural change in PAN crystals. In this work, it was found that CNTs facilitate the orientation of the PAN crystals during deformation (Figure 11) and increase the load-transferred to PAN crystals as evidenced by their increased lateral and axial strain at 75° C.

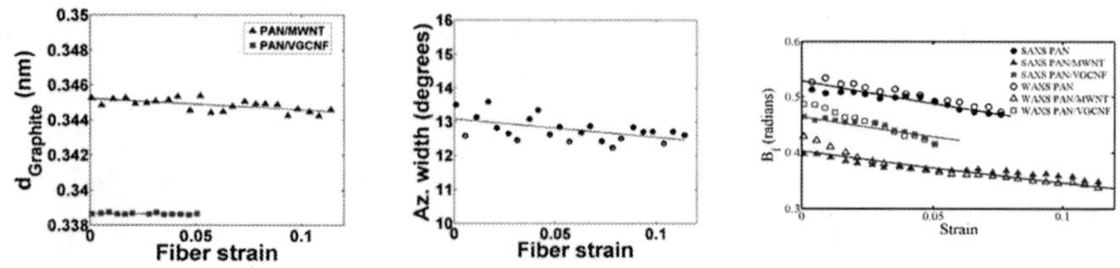

Figure 11.

a)-(c) from left to right. Changes in the *d*-spacing of carbon as a function of strain derived from the Bragg's peak of nanotubes at room temperature; (b) Degree of orientation of MWNT in the form of the azimuthal breadth.(Reprinted from Polymer, Vol.49, W. Wang, N.S. Murthy, H.G. Chae and S. Kumar, "Structural changes during deformation in carbon nanotube-reinforced polyacrylonitrile fibers", 2133-2145. Copyright 2008, with permission from Elsevier.) (c) Changes in the extent of misorientation (B_f) for scattering entities at room temperature. Black circle, blue triangle and red square represent PAN, PAN/MWNT, and PAN/VGCNF, respectively. Filled and empty symbols represent SAXS and WAXS results, respectively. Straight line segments are calculated based on affine deformation model. (Wang et al.,. 2009).

DEFORMATION OF THE CNTS IN THE COMPOSITE

The diffraction patterns shown in Figure 10 can be used to characterize the deformation of CNTs under stress by monitoring the changes in the (002) peak of CNTs. This interlayer distance d_{002} is used to ascertain the changes in the lattice dimension under strain (Figure 11). This figure also shows the changes in the azimuthal width of these reflections with strain as well as the crystallite size corresponding to this reflection. The changes in these parameters show that CNTs deform under load and thus contributes to the increased modulus, and also alter the response of the PAN matrix to stress, thus enhancing the performance of the composite.

CONCLUSIONS

Among the many techniques that are currently available, electron microscopy, Raman spectroscopy and x-ray scattering have proved to be most valuable to study the structure in polymer nanocomposites. These techniques complement each other by covering overlapping length scales from sub nm to μm. These techniques are valuable in assessing the dispersion, alignment of the nanotubes in the matrix, to probe the adhesion and the interface between the nanotube and the polymer, and to study the structural changes that occur due to the interaction between the two components as well as during deformation.

REFERENCES

1. S. G. Advani, 2006 Processing and Properties of Nanocomposites: World Scientific.
2. P. M. Ajayan, P. Redlich, M. Ruhle, 1997 Structure of carbon nanotube-based nanocomposites. Journal of Microscopy, 185 275282 .
3. P. M. Ajayan, L. S. Schadler, C. Giannaris, A. Rubio, 2000 Single-Walled Carbon Nanotube-Polymer Composites: Strength and Weakness. Advanced Materials, 12(10), 750-753.
4. C. Bower, R. Rosen, L. Jin, J. Han, O. Zhou, 1999 Deformation of carbon nanotubes in nanotube-polymer composites. Applied Physics Letters, 74(22), 3317-3319.

5. O. Breuer, U. Sundararaj, 2004 Big Return From Small Fibers: A Review of Polymer/Carbon Nanotube Composites. Polymer Composites, 25(6), 630-645.
6. A. Brosse, C. , S. Tence-Girault, P. M. Piccione, L. Leibler, 2008 Effect of multi-walled carbon nanotubes on the lamellae morphology of polyamide-6. Polymer, 49 46804686 .
7. P. Calvert, 1999 Nanotube composites: A recipe for strength. Nature, 399 210211 .
8. H. G. Chae, Y. H. Choi, M. L. Minus, S. Kumar, 2009 Carbon nanotube reinforced small diameter polyacrylonitrile based carbon fiber. Composites Science and Technology, 69 406413 .
9. H. G. Chae, S. Kumar, 2009 Making Strong Fibers. Science, 319(5865), 908-909.
10. H. G. Chae, M. L. Minus, S. Kumar, 2006 Oriented and exfoliated single wall nanotubes in polyacrylonitrile. Polymer, 47(10), 3494-3504.
11. H. G. Chae, M. L. Minus, A. Rasheed, S. Kumar, 2007 Stabilization and Carbonization of gel spun polyacrylonitrile/single Wall nanotube Composite Fibers. Polymer, 48 37813789 .
12. H. G. Chae, T. V. Sreekumar, T. Uchida, S. Kumar, 2005 A comparison of reinforcement efficiency of various types of carbon nanotubes in polyacrylonitrile fiber. Polymer, 46 1092510935 .
13. T. Chatterjee, C. A. Mitchell, V. G. Hadjiev, R. Krishnamoorti, 2007 Hierarchical Polymer-Nanotube Composites. Advanced Materials, 19 38503853 .
14. X. Chen, C. Burger, D. Fang, I. Sics, X. Wang, W. He, et al. 2006 In-Situ X-ray Deformation Study of Fluorinated Multiwalled Carbon Nanotube and Fluorinated Ethylene-Propylene Nanocomposite Fibers. Macromolecules, 39 54275437 .
15. J. N. Coleman, U. Khan, Y. K. Gun'ko, 2006 Mechanical Reinforcement of Polymers Using Carbon Nanotubes. Advanced Materials, 18 689705 .
16. C. A. Cooper, S. R. Cohen, A. H. Barber, H. D. Wagner, 2002 Detachment of nanotubes from a polymer matrix. Applied Physics Letters, 81(20), 3873-3875.
17. V. P. Dravid, X. Lin, Y. Wang, X. K. Wang, A. Yee, J. B. Ketterson, et al. 1993 Buckytubes and Derivatives: Their Growth and Implications for Buckyball Formation. Science, 259(5101), 1601-1604.
18. M. S. Dresselhaus, G. Dresselhaus, R. Saito, A. Jorio, 2005 Raman spectroscopy of carbon nanotubes. Physics Reports, 409 4799 .
19. F. T. Fisher, R. D. Bradshaw, L. C. Brinson, 2002 Effects of nanotube

waviness on the modulus of nanotube-rein. Applied Physics Letters, 80, 4647--4649.

20. S. J. V. Frankland, A. Caglar, D. W. Brenner, M. Griebel, 2002 Molecular Simulation of the Influence of Chemical Cross-Links on the Shear Strength of Carbon Nanotube-Polymer Interfaces. J. Phys. Chem. B, 106 30463048 .
21. M. Ge, K. Sattler, 1993 Vapor-Condensation Generation and STM Analysis of Fullerene Tubes. Science, 260(5107), 515-518.
22. H. H. Gommans, J. W. Alldredge, H. Tashiro, J. Park, J. Magnuson, A. G. Rinzler, 2000 Fibers of aligned single-walled carbon nanotubes: Polarized Raman spectroscopy. Journal of Appied Physics, 88 25092514 .
23. B. P. Grady, F. Pompeo, R. L. Shambaugh, D. E. Resasco, 2002 Nucleation of Polypropylene Crystallization by Single-Walled Carbon Nanotubes. Journal of Physical Chemistry B, 106 58525858 .
24. H. Guo, T. V. Sreekumar, T. Liu, M. Minus, S. Kumar, 2005 Structure and Properties of Polyacrylonitrile/single wall carbon nanotube composite films. Polymer, 46 30013005 .
25. Y. Y. Huang, S. V. Ahir, E. M. Terentev, 2006 Dispersion rheology of carbon nanotubes in a polymer matrix. Physical Review B, 73, 125422.
26. S. Iijima, 1991 Helical microtubules of graphitic carbon. Nature, 354 5658 .
27. T. Inada, H. Masunaga, S. Kawasaki, M. Yamada, K. Kobori, K. Sakurai, 2005 Small-angle X-ray Scattering from Multi-walled Carbon Nanotubes (CNTs) Dispersed in Polymeric Matrix. Chemistry Letters, 34(4), 524-525.
28. A. Koganemaru, Y. Bin, Y. Agari, M. Matsuo, 2004 Composites of Polyacrylonitrile and Multiwalled Carbon Nanotubes Prepared by Gelation/Crystallization from solution. Advanced Functional Materials, 14(9), 842-850.
29. T. Liu, S. Kumar, 2003 Quantitative characterization of SWNT orientation by polarized Raman spectroscopy. Chemical Physics Letters, 378 257262 .
30. T. Liu, I. Y. Phang, L. Shen, S. Y. Chow, W. Zhang, D. , 2004 Morphology and Mechanical Properties of Multiwalled Carbon Nanotubes Reinforced Nylon-6 Composites. Macromolecules, 37 72147222 .
31. Y. Maniwa, R. Fujiwara, H. Kira, H. Tou, E. Nishibori, M. Takata, et al. 2001 Multiwall carbon nanotubes grown in hydrogen atmosphere: An X-ray study. Physical Review B, 64 073105073104 .
32. G. H. Michler, 2008 Electron Microscopy of Polymers (1st ed.): Springer.

33. M. Moniruzzaman, K. I. Winey, 2006 Polymer Nanocomposites Containing Carbon Nanotubes. Macromolecules, 39 51945205 .
34. M. i. h. Panhuis, A. Maiti, A. B. Dalton, A. v. d. Noort, J. N. Coleman, B. Mc Carthy, et al. 2003 Selective Interaction in a Polymer-Single-Wall Carbon Nanotube Composite. J. Phys. Chem. B, 107 478482 .
35. E. Pasqualini, 1997 Concentric carbon structure. Physical Review B, 56(13), 7751-7754.
36. V. Pichot, S. Badaire, P. A. Albouy, C. Zakri, P. Poulin, P. Launois, 2006 Structural and mechanical properties of single-wall carbon nanotube fibers. Physical Review B, 74(245416), 245416.
37. D. Qian, E. C. Dickey, R. Andrews, T. Rantell, 2000 Load Transfer and deformation mechanisms in carbon nanotube-polystyrene composites. Applied Physics Letters, 76(20), 2868-2870.
38. A. M. Rao, A. Jorio, M. A. Pimenta, M. S. S. Dantas, R. Saito, G. Dresselhaus, et al. 2000 Polarized Raman Study of Aligned Multiwalled Carbon Nanotubes. Physical Review Letters, 84 18201823 .
39. L. Sawyer, D. T. Grubb, G. F. Meyers, 2010 Polymer Microscopy (3rd ed.): Springer.
40. L. Schadler, 2007 Nanocomposites: Model Interface. Nature Materials, 6 257258 .
41. D. W. Schaefer, J. M. Brown, D. P. Anderson, J. Zhao, K. Chokalingam, D. Tomlin, et al. 2003a Structure and dispersion of carbon nanotubes. J. Appl. Cryst., 36 553557 .
42. D. W. Schaefer, J. Zhao, J. M. Brown, D. P. Anderson, D. W. Tomlin, 2003b Morphology of dispersed carbon single-walled nanotubes. Chemical Physics Letters, 375 369375 .
43. M. S. P. Shaffer, A. H. Windle, 1999 Fabrication and characterization of carbon nanotube/poly(vinyl alcohol) composites. Advanced Materials, 11 937941 .
44. T. V. Sreekumar, T. Liu, B. G. Min, H. Guo, S. Kumar, R. H. Hauge, et al. 2004 Polyacrylonitrile Single-Walled Carbon Nanotube Composite Fibers. Advanced Materials, 16(1), 58-61.
45. D. Srivastava, C. Wei, K. Cho, 2003 Nanomechanics of carbon nanotube and composite. Appl. Mech. Rev 56(2), 215-230.
46. E. T. Thostenson, T. Chou, W. , 2002 Aligned multi-walled carbon nanotube-reinforced composites: processing and mechanical characterization. J. Phys. D: Appl. Phys., 35, L77 -L80.
47. M. M. J. Treacy, T. W. Ebbesen, J. M. Gibson, 1996 Exceptionally high Young's modulus observed for individual carbon nanotubes. Nature, 381 678680 .

48. T. Uchida, D. Anderson, M. Minus, S. Kumar, 2006 Morphology and modulus of vapor grown carbon nano fibers Journal of Materials Science, 41(18), 5851-5856.
49. W. Wang, P. Ciselli, E. Kuznetsov, T. Peijs, A. H. Barber, 2008a Effective reinforcement in carbon nanotube-polymer composites. Philosophical Transactions of the Royal Society A: Mathematical, Physical and Engineering Sciences, 366 16131626 .
50. W. Wang, N. S. Murthy, H. G. Chae, S. Kumar, 2008b Structural changes during deformation in carbon nanotube-reinforced polyacrylonitrile fibers. Polymer, 49(8), 2133-2145.
51. W. Wang, N. S. Murthy, H. G. Chae, S. Kumar, 2009 Small-Angle X-ray Scattering Investigation of Carbon Nanotube-Reinforced Polyacrylonitrile Fibers During Deformation. Journal of Polymer Science: Part B: Polymer Physics.
52. M. Wong, M. Paramsothy, X. J. Xu, Y. Ren, S. Li, K. Liao, 2003 Physical interactions at carbon nanotube-polymer interface. Polymer, 44 77577764 .
53. G. Xu, Z.-c. Feng, Z. Popovic, J.-y. Lin, J. J. Vittal, 2001 Nanotube structure Revealed by High-resolution X-ray Diffraction. Advanced Materials, 13(4), 264-267.
54. H. Ye, H. Lam, T. Nick, Y. Gogotsi, K. Frank, 2004 Reinforcement and rupture behavior of carbon nanotubes-polymer nanofibers. Applied Physics Letters, 85(10), 1775-1777.
55. Y. C. Zhang, X. Wang, 2005 Thermal effects on interfacial stress transfer characteristics of carbon nanotubes/polymer composites. International journal of solids and structures, 42 53995412 .
56. O. Zhou, R. M. Fleming, D. W. Murphy, C. H. Chen, R. C. Haddon, A. P. Ramirez, et al. 1994 Defects in Carbon Nanostructure. Science, 263 17441747 .

Chapter 14

ELECTRICAL PROPERTIES OF CNT-BASED POLYMERIC MATRIX NANOCOMPOSITES

Alessandro Chiolerio[1], Micaela Castellino[1], Pravin Jagdale[2], Mauro Giorcelli[2], Stefano Bianco[3] and Alberto Tagliaferro[2]

[1] Physics Department, Politecnico di Torino, Italy

[2] Materials Science and Chemical Engineering Department, Politecnico di Torino, Italy

[3] Fondazione IIT (Istituto Italiano di Tecnologia), Centre for Space Human Robotics Torino, Italy

INTRODUCTION

NanoComposites (NCs) are a class of materials widely investigated in the last decade. The term NC material has broadened significantly to encompass a large variety of systems such as one-dimensional, two-dimensional, three-dimensional and amorphous materials, made of distinctly dissimilar components and mixed at the nanometer scale. The general class of organic/inorganic NC materials is a fast growing area of research. The properties of NCs depend not only on the properties of their individual elements but also on their morphology and interfacial characteristics. Large interface area between the matrix and the nano filler is a key issue for NCs.

NCs' fillers include Carbon nanomaterials. They are a large family of materials carbon based that include: fibers, nanotubes, fullerene, nano-diamonds, etc. Carbon Nanotubes (CNTs) are one of the most popular and intensively studied Carbon nanomaterials.

Since their discovery in 1991 by Iijima (Iijima, 1991), they have attracted great interest as an innovative material in most fields of science and engineering. CNTs can be thought of as sheets of graphite rolled up to make a tube. They are divided in two large classes: Multi wall CNTs (MWCNTs) and Single wall CNTs (SWCNTs). This classification depends on the number of graphite walls: several in the case of MWCNTs and only one for SWCNTs. The dimensions are variable, from few nanometers for SWCNTs to tenths of nanometers for MWCNTs. CNTs have outstanding mechanical, thermal and electrical properties.

For example as shown by Collins et al. (Collins et al., 2000) CNTs have electrical properties and electric-current-carrying capacity 1000 times higher than a copper wire. Young's module for a single-walled carbon nanotube (SWCNTs) has been estimated by Yu et al. (Yu et al., 2000) in a range of 0.32-1.47 TPa and strengths between 10 and 52 GPa with a toughness of ~ 770 Jg^{-1}. A room-temperature thermal conductivity of 1750–5800 $Wm^{-1}K^{-1}$ has been estimated by Hone et al. (Hone et al., 1999).

These remarkable properties make them excellent candidates for a range of possible new classes of materials.

Several studies demonstrate how just a small percentage of CNTs loading can improve the material properties (Breuer, 2004), while still maintaining the plasticity of the polymers.

The most accessible near-term application for CNTs-polymer composite involves their electrical properties. The intrinsic high conductivity of CNTs makes them a logical choice for tuning the conductive properties of polymers. As demonstrated by MacDiarmid (MacDiarmid, 2002) CNT-added polymers are ideal candidate because they can increase the electrical conductivity by many orders of magnitude from 10^{-10} – 10^{-5} up to 10^{3} – 10^{5} Scm^{-1}.

CNTs production cost depends upon several parameters, one of them is the type of CNTs. In general the production of SWCNTs is more expensive than MWCNTs due to low quantity production. Purification, surface modification treatments and functionalisation also increase the cost of CNTs production. For commercialization of process it is important to use cost effective and easily reproducible CNTs i.e. MWCNTs, whose diffusion, in the last years, seems to be greater.

Depending on the specific application of CNT-based NCs, it is possible to create either an isotropic material or an anisotropic one, by orienting preferentially the CNTs: its physical properties will be in the first case independent of sample positioning and geometry, in the second case strongly geometry-dependent (Chiolerio et al., 2008). In the case of NCs prepared for electrical applications, it is desirable to create a homogeneous polymer composite.

Carboxylic functionalized CNTs (-COOH groups) for their well known easy dispersion in polymers (Gao et al., 2009) are the ideal candidate in polymer composites. In fact the modification of their surface decreases their hydrophobic nature and improves interfacial adhesion to a bulk polymer through chemical attachment. Easy and good dispersion of CNTs in the polymer matrix is essential to obtain a homogeneous final product. Small diameter CNTs, providing a uniform dispersion, means more CNTs per volume unit and as a consequence a more capillary distribution in the polymer matrix at the same weight percentage. Following these reasoning the choice of CNTs has been done on a commercial product (Nanocyl™ NC3101) ready to our purposes (see all details in experimental section).

The choice of the polymers has focused on low cost materials, commercial products featuring good affinity with selected CNTs. A thermoplastic polyolefine: polyvinyl butyral (PVB) normally used in the area of glass gluing. A syloxane, Polydimethylsiloxane (PDMS) the most widely silicon-based organic polymer used. Two thermoset commercial epoxy resins: Epilox™ used in the automotive field (E) and Henkel Resin Hysol EA-9360 leader in the aerospace field (H1).

In this chapter a detailed study is presented, on the experimental conductivity and scaling laws thereof, for each of the four polymeric matrices above mentioned, as a function of the CNTs volume fraction. It will be shown that, depending on the polymer choice, it is possible to obtain NCs having a DC conductivity either showing a huge dependence on the volume fraction of CNTs or a less marked one. We also found that in one case there is no typical conductivity kink in correspondence of the percolation threshold, for the PVB matrix.

EXPERIMENTAL

COMPOSITE PROCESSING

The effective utilization of CNT materials in composite applications depends strongly on their ability to be dispersed individually and homogeneously within a matrix. To maximize the advantage of CNTs as effective reinforcement for high strength polymer composites, the CNTs should not form aggregates, and must be well dispersed to enhance the interfacial interaction with the matrix. Several processing methods are available for fabricating CNT/polymer composites. Some of them include solution mixing, in situ polymerization, melt blending and chemical modification processes.

The general protocol for solution processing method includes the dispersion of CNTs in a liquid medium by vigorous stirring and sonication, mixing the CNTs dispersion solvents in a polymer solution and controlled evaporation of the solvent with or without vacuum conditions.

In experimental part, solution processing method was adopted for making polymer NCs.

PVB and MWCNTs were treated with solvent like ethanol and butanol. Commercial thermosetting resins E, H1 and PDMS with MWCNTs followed the same method without solvent due to their liquid/viscous nature. In this way, a satisfactory level of dispersion of CNTs into the polymer matrix was obtained. Different weight % (wt.-%) concentrations of MWCNTs in polymer resin were tried to study the electrical behavior of the polymer NC.

CHARACTERIZATION OF FUNCTIONALIZED MWCNTS

Commercial NC3101 MWCNTs (NANOCYL™) are produced via the catalytic carbon vapour deposition (CCVD) process with average diameter 9.5 nm and average length ~ 1.5 μm (Figure 1, Field Effect Secondary Electron Microscope, FESEM).

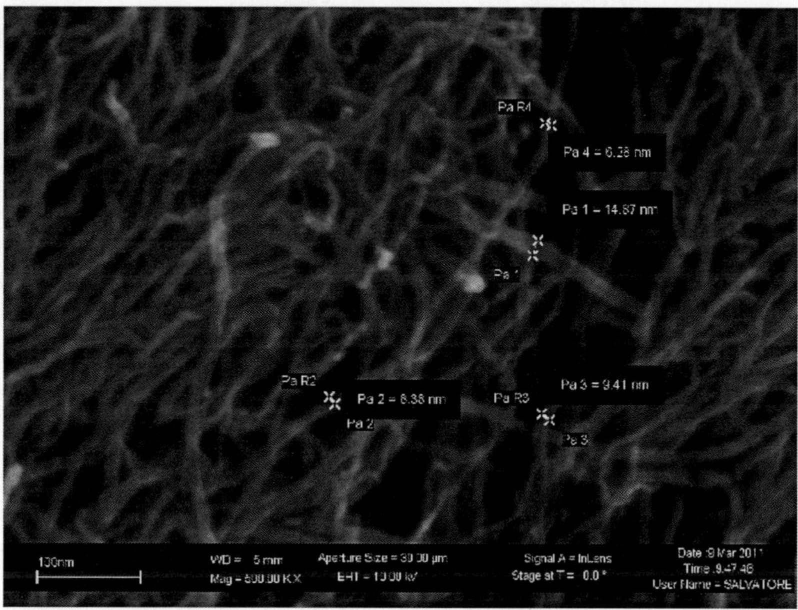

Figure 1. FESEM image of Nanocyl™ (NC3101) Functionalized MW-CNTs.

MWCNTs are then purified to greater than 95% carbon to produce the 3100 grade. This product is then functionalized with a carboxylic group (-COOH) to produce the 3101 grade. This functionalization is useful to avoid the clumps and agglomeration of MWCNTs in Polymer resin.

THERMOSET EPOXY RESIN (E)

RESIN (T 19-36/700)

It is a commercially modified, colorless, low viscosity (650-750 mPa·s at 25° C) epoxy resin with reduced crystallization tendency having density 1.14 gcm^{-3}. Cross linking with suitable curing agents is preferably performed at room temperature. The chemical composition of Epilox resin T19-36/700 is mainly Bisphenol A (30 – 60 wt.-%), added by crystalline silica (quartz) (1 – 10 wt.-%), glycidyl ether (1 – 10 wt.-%), inert fillers (10 –60 wt.-%).

HARDENER (H 10-31)

It is a liquid, colourless, low viscosity (400-600 mPa.s) modified cycloaliphatic polyamine epoxide adduct having density 1 gcm^{-3}. Modified cycloaliphatic amine adducts.

SAMPLES PREPARATION

Resin T 19-36/700, MWCNTs and hardener H 10-31 were mixed with vigorous stirring followed by sonication. The mixture was degassed in vacuum. The mixture was poured in a mold to acquire desired shape for electrical measurements and subsequently kept in an oven at 70° C for 4 hours for curing purposes. Electron microscopy images show the dispersion of MWCNTs in E (Figure 2).

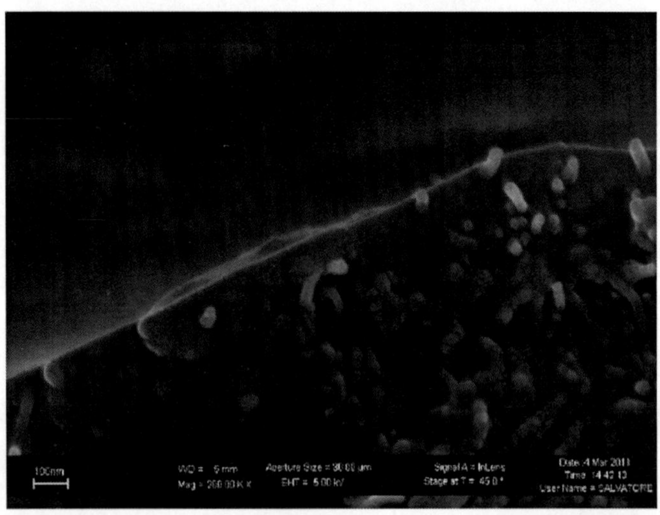

Figure 2. FESEM of E + 5 wt.-% of MWCNTs.

THERMOSET HENKEL RESIN (H1)

HYSOL EA-9360 PART A

Henkel (Hysol EA-9360 Part A) is a white viscous paste having density 1.18 gml^{-1}. It has some interesting properties like high peel strength, excellent static stress durability, room temperature curability.

HYSOL EA-9360 PART B

Henkel (Hysol EA-9360 Part B) is a blue paste having density 1 gml^{-1}. The mixing ratio of Part A and Part B is 100:43 by weight. The chemical composition of the Henkel Part A and Part B is given below (Table 1).

SAMPLE PREPARATION

Combining Part A, Part B and MWCNTs in the correct ratio and mixing thoroughly with stirring was performed. Subsequently: degassing the mixture in vacuum, mixing of Part A and Part B which resulted in exothermic reaction (mixing smaller quantities would minimize the heat build up). The bonded parts were held in contact until the adhesive was set. Handling strength for this adhesive occurs in 24 hours (>25° C) and complete curing achieves after

Henkel Resin Hysol (EA-9360) Chemical Composition (wt.-%)			
Part A		Part B	
Epoxy resin Proprietary	30-60	Piperazine derivative Proprietary	30-60
Polyfunctional epoxy resin Proprietary	10-30	Butadiene-acrylonitrile copolymer	10-30
Synthetic rubber Proprietary	10-30	Silica amorphous (fumed)	5-10

Glass spheres Proprietary	5-10	Benzyl alcohol	5-10
Filler Proprietary	1-5	Cycloaliphatic amine Proprietary	5-10
Substituted silane Proprietary	1-5	Phenol	1-5
Substituted Piperazine Proprietary		Diethylene glycol Di-(3-aminopropyl)ether	1-5
		1-5	

Table 1. Chemical composition of Henkel Part A and Part B.

5-7 days keeping at 25° C, for faster curing the molds were kept in the oven at 82° C for 1 hour. Electron microscopy image shows the dispersion of MWCNTs in H1 (Figure 3).

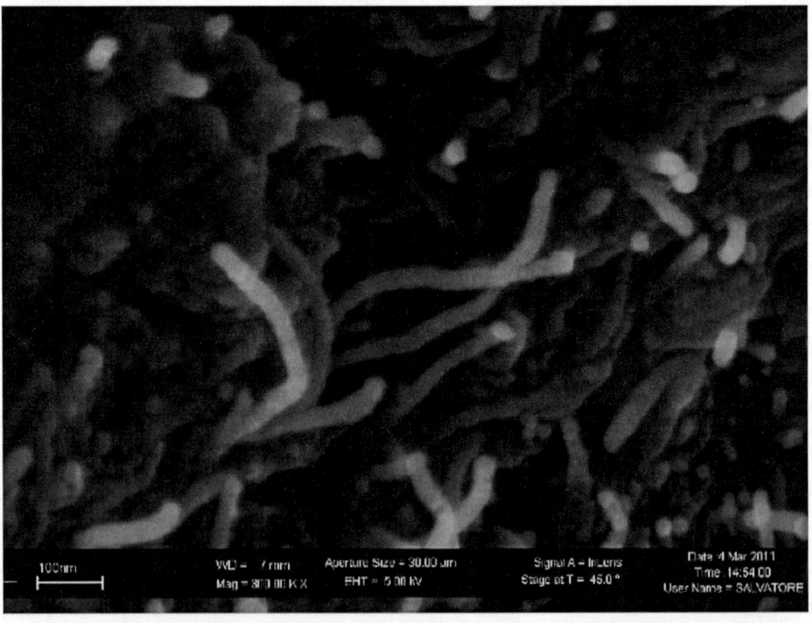

Figure 3. FESEM of H1+ 5 wt.-% of MWCNTs.

POLY (DIMETHYLSILOXANE) (PDMS)

BASE (SILICON ELASTOMER SYLGARD 184 (DOW CORNING))

It is a Silicon based clear colourless low viscous liquid having specific gravity 1.11 gcm^3. It is chemically stable and not forming hazardous polymerization.

CURING AGENT (SILICON ELASTOMER SYLGARD 184 (DOW CORNING))

It is silicon resin clear colourless low viscous liquid having specific gravity 1.03 gcm^{-3}.It is also chemically stable and not forming hazardous polymerization like base. The mixing ratio of base and curing agent is 1:1 by weight and curing time for the composite is 48 hours at 25° C. Chemical formula of PDMS is CH_3 $[Si\ (CH_3)_2O]_n$ $Si(CH_3)_3$. After polymerization and cross-linking, PDMS samples present an external hydrophobic surface.

METHOD FOR SAMPLE PREPARATION

To ensure uniform distribution of MWCNTs, Base and Curing agent must be thoroughly mixed prior to their combination in a 1:1 ratio. The mixture was stirred and sonicated well followed by degassing in vacuum. The base and curing agent liquid mixture should have a uniform appearance. The presence of light-colored streaks or marbling indicates inadequate mixing and will result in incomplete cure. Electron microscopy image shows the dispersion of MWCNTs in PDMS resin (Figure 4).

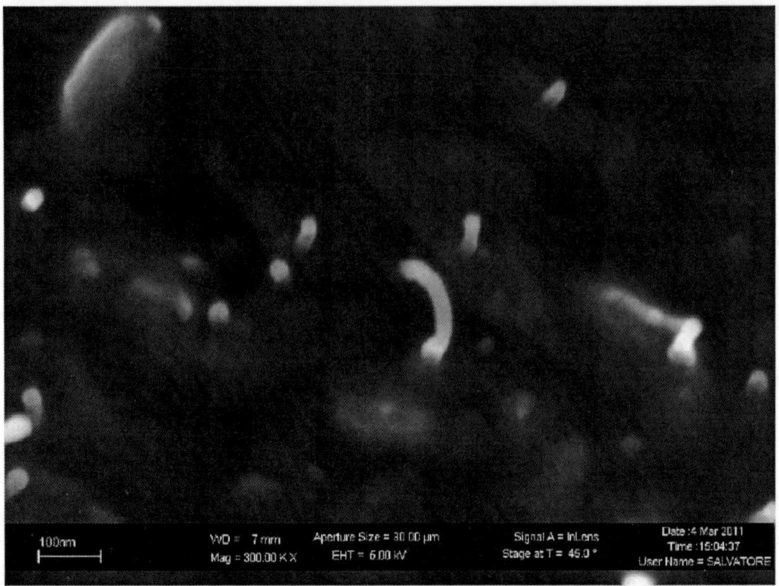

Figure 4. FESEM of PDMS and 5 wt.-% MWCNTs.

POLY (VINYL BUTYRAL) (PVB)

PVB is a resin usually used for applications that require strong binding, optical clarity, adhesion to many surfaces, toughness and flexibility. It is prepared from polyvinyl alcohol by reaction with butyraldehyde. The IUPAC name of the polymer is Poly[(2-propyl-1,3-dioxane-4,6-diyl) methylene] with chemical structure as mentioned below (Figure 5). It is in white powder form with specific gravity 1.083 gcm^{-3}.

METHOD OF PREPARATION OF PVB/MWCNTS NCS BY SOLUTION PROCESSING

In preparation of CNTs–PVB (Butvar B-98, Sigma Aldrich) polymer composites, Ethanol (Carlo Erba) and 1-Butanol (Sigma-Aldrich) solvents were used with vigorous stirring and sonication. Degassing was important for eliminating the entrapped solvent gas bubbles under vacuum. The composite was treated in oven at 70° C for curing.

Electron microscopy images show the dispersion of MWCNTs in PVB (Figure 5).

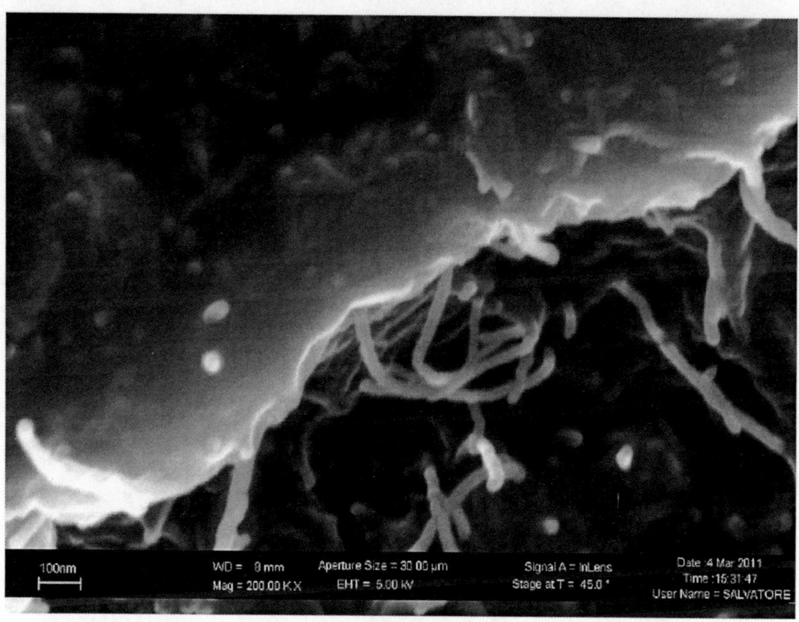

Figure 5. PVB with 5 wt.-% of MWCNTs.

ELECTRICAL CHARACTERIZATION

Electrical measurements were performed for each sample of the four polymer series using the so called *"Two Point Probe (TPP) method"* (Schroder, 1990)(Grove, 1993). Since we have dealt with, in principle, highly non conductive matrices, we have decided not to use the in-line *"Four Point Probes (FPP) technique"* (Smith, 1958), which is typically used for highly conductive materials, where the contact resistance between sample and connection wires could play an important role, leading to a wrong estimate of the real material resistance (Chiolerio et al., 2007).

In order to perform I-V measurements a setup already tested was chosen and used for the estimate of CNTs bundles resistivity at room temperature (Castellino et al., 2010).

In this case it was adjusted to fit our samples' dimensions. As it is possible to see from the scheme inFigure 6 a square sample was placed onto an insulating board with tin islands to place electrical connections for the Voltage supplier. To create the connections directly onto our samples surface, in an easy, fast and reproducible way, the silver conductive paste was chosen (Hu et al., 2008), which does not require any particular equipment (such as metals sputtering or thermal evaporators devices) (Coleman et al., 1998), just several minutes in order to let the solvent to evaporate. In this way it was also prevented any kind of warming of NCs or metals infiltrations during the metal contacts creation, which could have changed their properties.

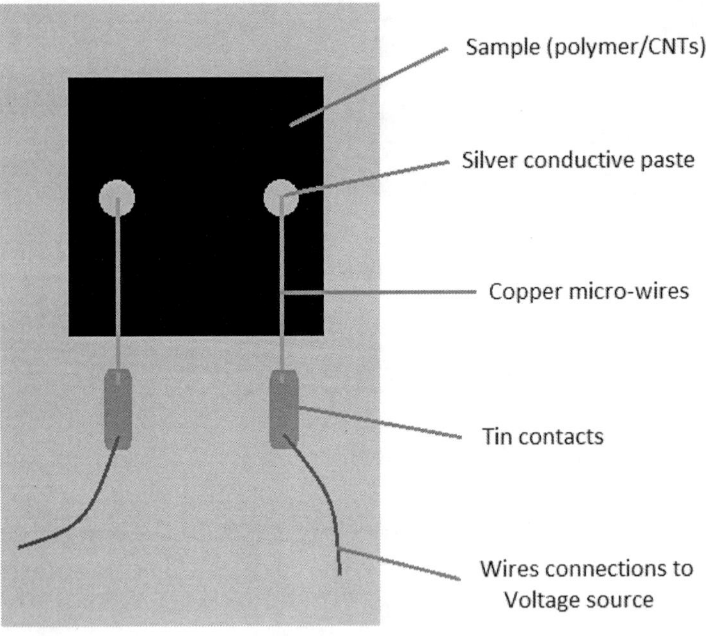

Figure 6. Scheme of the electrical circuit device used for I-V measurements.

Insulated copper micro-wires were used to connect the Ag paste onto the sample and the tin islands on the board, since they are very thin and can be easily attached with this conductive paste. Their ends were previously exposed by means of a scalpel under an optical microscope.

A Keithley-238 High Current Source Measure Unit was used as high voltage source and nano-amperometer. It can supply up to ± 110 V and measure up to ±1 A (at ±15 V) and down to 10 fA.

Each sample was measured three times to have a statistical average value for the resistance and to see if the behavior observed was reproducible. For this purpose electrical connections between the voltage supplier and the board were removed every time before each measure.

For samples with 0 wt.-% CNTs the I-V measurements were performed in the range between ±110 V with a voltage step of 1 V, in order to observe a sort of trade in the very noisy curve where the current was $I \sim 10^{-11} - 10^{-10}$ A (see Figure 7 upper panel) to obtain a sort of "*zero value*" for the conduction (σ_0), which will be used further in the results and discussion section for the percolation theory fit.

Instead a narrower range (±10 V or ± 20V, with a step of 0.1 V) was chosen for samples with wt.-% CNTs ≥ 1 mainly for two reasons:

- to avoid samples warming due to high current flow;
- since the voltage supplier possesses a threshold cut-off for currents higher than 1 A.

The first phenomenon has been observed especially with samples with a CNTs load equal or higher than 4 wt.-%. For this reason the range was moreover reduced between ±7 or ± 3 V for some highly conductive samples (see Table 2 and Figure 7 lower panel).

In fact for E and PDMS composites a huge warming of the samples was observed, especially for PDMS + 5 wt.-% CNTs, which produced fumes from the silver paste.

On the other hand I1 NCs showed quite standard behaviors, showing an increase in the conductivity due to the increase in the CNTs amount, with no remarkable aspects to be noticed.

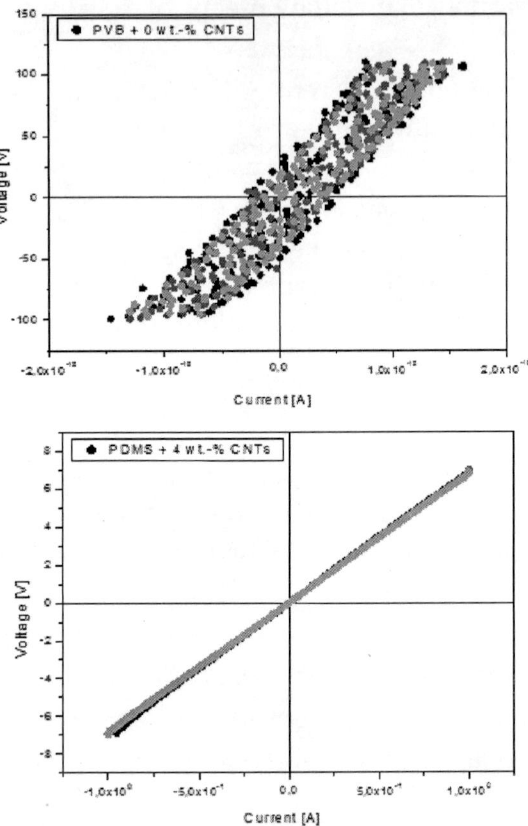

Figure 7.

Noise curves of samples PVB + 0 wt.-% CNTs (upper panel) and reduced voltage range measurements for sample PDMS + 4 wt.-% CNTs (lower panel).

Table 2.

Polymer	wt.-% CNTs	Voltage range (± V)	Voltage step (V)	Current range (± A)
E	0	110	1	1×10-10
	1	10	0.1	2×10-4

	2	10	0.1	$3\times1^{0-}2$
	3	10	0.1	$3\times1^{0-}1$
	4	10	0.1	$8\times1^{0-}1$
	5	7	0.1	$8\times1^{0-}1$
H1	0	110	1	$2\times1^{0-1}0$
	1	110	1	$1\times1^{0-1}0$
	2	10	0.1	$1.5\times1^{0-}6$
	3	10	0.1	$3\times1^{0-}4$
	4	10	0.1	$1.5\times1^{0-}2$
	5	10	0.1	$4\times1^{0-}2$
PDMS	0	110	1	$1\times1^{0-1}0$
	1	10	0.1	$1\times1^{0-}3$
	2	10	0.1	$1\times1^{0-}1$
	3	10	0.1	$5\times1^{0-}1$
	4	7	0.1	$1\times1^{0}0$
	5	3	0.1	$4\times1^{0-}1$
PVB	0	110	1	$1.5\times1^{0-}$ $^{1}0$
	1	10	0.1	$2.5\times1^{0-}$ $^{1}0$
	2	10	0.1	$8\times1^{0-}5$
	3	20	0.1	$6\times1^{0-}5$
	4	10	0.1	$2\times1^{0-}3$
	5	10	0.1	$6\times1^{0-}3$

Voltage (supplied) and current (registered) ranges for each sample.

RESULTS AND DISCUSSION

FINITE ELEMENT METHOD SIMULATION

Electrical resistivity measures were performed on the samples produced as described in the previous section. Figure 8 shows the sample geometry and the layout of the silver electrodes, placed in the 2-point configuration.

A Finite Element Method simulation was performed, using the commercial code Comsol Multiphysics™, of a composite material slab characterized by different resistivities, having the same dimensions of real samples, with the aim of evaluating the volume interested by the higher fraction of current density and estimating the penetration depth of DC currents into the sample thickness. An example of the simulation control volume is given in Figure 9, where the tetrahedral mesh of Lagrangian cubic elements is shown. The complexity of the system was quite high, having 162 k degrees of freedom and 111 k elements composing the

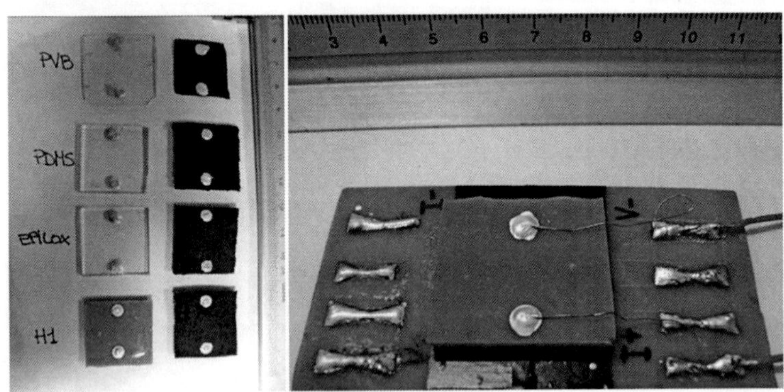

Figure 8.

Samples ready for electrical measurements; left panel: left column, pure matrices; right column, composites containing 5 wt.-% of CNTs.

Right panel: sample placed in the fixture used for DC measurements. mesh, with a solution time of more than 315 s (Intel™ Core® 2 Quad Q9550 2.83 GHz 4 GB DDR3).

As it can be seen in Figure 10, the current density is distributed almost in the whole sample, with the exception of the portions close to the electrodes, where the effective path avoids the sample bottom and edges. Based on these simulations, the effective electrical path was estimated to be: 3 mm thick (same thickness of the sample), 3 cm width (same width of the sample) and 1 cm long (sample length reduced by the electrode size and dead ends).

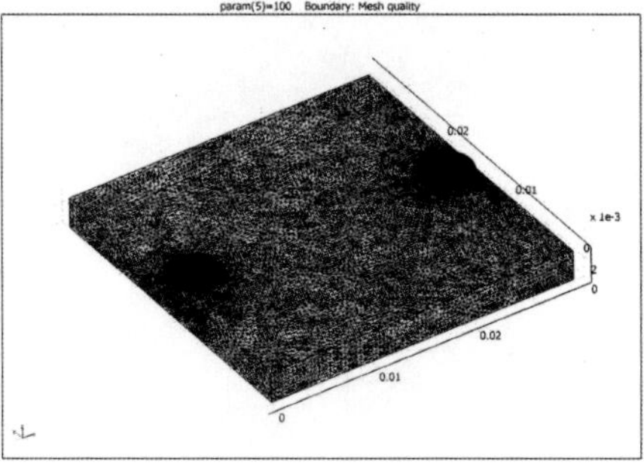

Figure 9. Mesh distribution on the control volume of the FEM simulation.

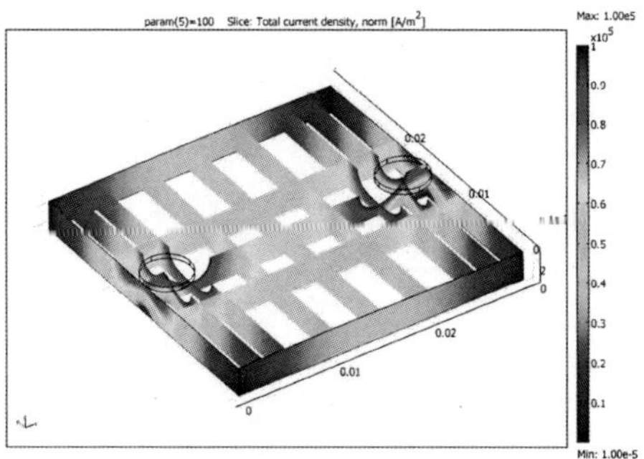

Figure 10. Current density distribution in a composite volume.

DC CONDUCTIVITY AND SCALING LAWS THEREOF

NCs samples showed two different electrical behaviors: linear response and non-linear response, in the voltage range ±10 V. In particular, samples characterized by the PDMS, PVB and E matrices, at low dispersoid concentrations, resulted in a symmetric nonlinear response, as shown in Figure 11. In particular, at least two regimes were observed, that typical of PDMS composites, having higher conductivities (mA range under 10 V) and that typical of PVB composites, having much lower conductivities (μA range under 10 V). The same samples, by increasing the amount of CNTs, resulted in a perfect linear response (Figure 12), as well as H1 samples.

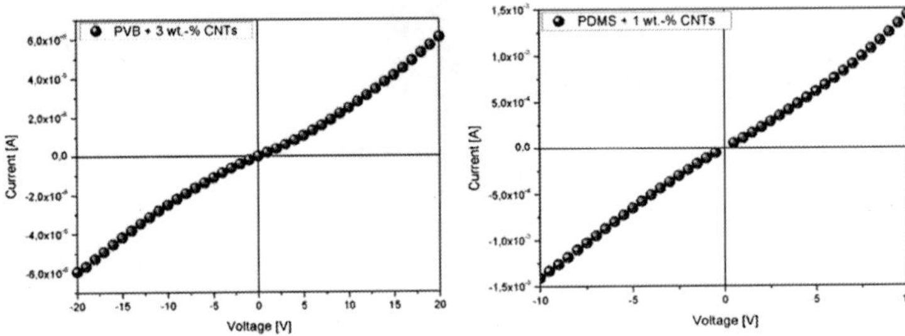

Figure 11. Nonlinear response of samples PVB + 3 wt.-% CNTs (left panel) and PDMS + 1 wt.-% CNTs (right panel). For clarity, only 1 experimental point every 5 is shown.

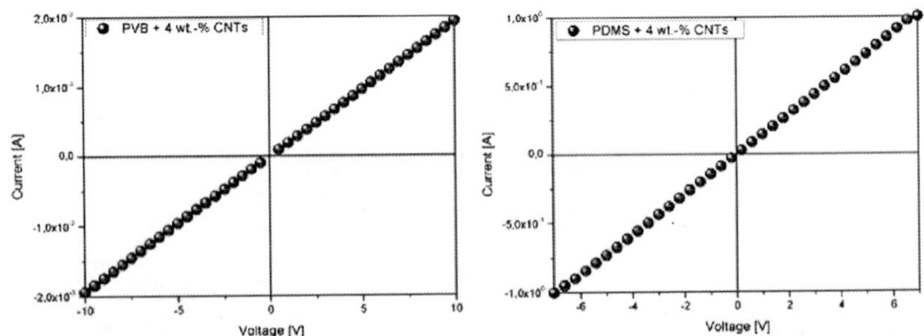

Figure 12.

Linear response of samples PVB + 4 wt.-% CNTs (left panel) and PDMS + 4 wt.-% CNTs (right panel). For clarity, only 1 experimental point every 5 is shown.

In literature, a nonlinear behavior was observed in composites based on MWCNTs networks dispersed in PDMS rubber even though the authors report no transition towards a linear response by increasing the amount of dispersoid (Liu and Fan 2007).

A possible explanation is given by the fluctuation-induced tunnelling mechanism acting in correspondence of CNTs' intersections. Some intersections, occurring between a metallic and a semiconductive CNTs, result in a Schottky barrier; when the dispersoid amount is low enough that no percolative random walks are formed going through point contact intersections, Schottky-like intersections play a major role and influence the macroscopic behavior of the NC. The energy necessary for the carriers to tunnel across the potential barrier consistent with a certain amount of polymer spacer placed between two CNTs is given by local temperature oscillations (Sheng, 1978).

In order to model the results obtained for conductivity for the different NCs as a function of the dispersoid amount, percolation theory was used as universal framework to describe physical properties of disordered systems (Stauffer & Aharony, 1994). Conductivity was computed by fitting the experimental data at high field values, thus neglecting eventual nonlinear effects which appear in the low voltage region. The idea behind the theory is that, given a lattice of a certain dimensionality and coordination, representing the dissipative or insulating matrix, a certain concentration of conductive nodes is randomly added to the system. By increasing the number of conductive nodes beyond a certain limit, said percolation threshold, a complete random walk across the lattice is made available to the carriers and a conductive behavior is observed.

$\sigma \infty (p-pc)t$

Equation (1) indicates that conductivity σ is proportional to a quantity depending on the volume fraction p of conductive dispersoid, on the percolation threshold p_c and on a universal scaling exponent t. Figure 13 shows the fit to experimental data, relative to our samples. The extrapolated parameters are expressed in Table 3. Generally the model seems unsuitable to fit experimental data relative to the PVB

NCs: no clear transition between insulating and conductive behavior is seen and the trend of conductivity versus CNTs volume fraction is linear, in semi-logarithmic scale (Figure 13, left panel). Remarkable is the case of matrix H1, showing an insulating state even when added by more than 1 v.-% of CNTs. E and PDMS matrices, for comparison, present a clear transition. Referring to Table 3data, we can see that the lowest percolation threshold was found for E resin (0.2±0.1 v.-%), followed by PDMS (0.7±0.2 v.-%). The critical exponent t is always close to 2.0, which is the expected value for a 3-dimensional sample (Kilbride et al., 2002).

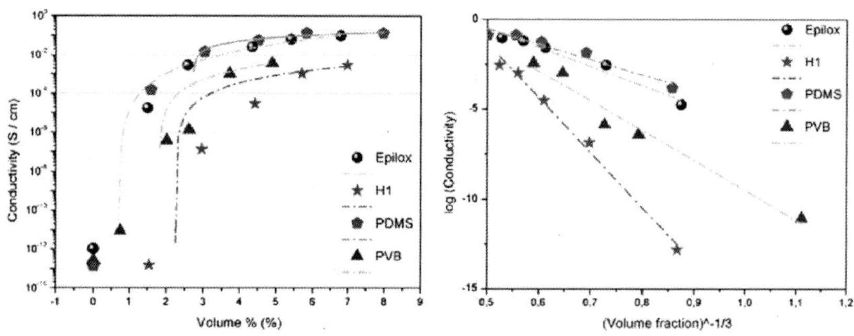

Figure 13.

Left panel: DC conductivity as a function of dispersoid volume fraction fitted by percolation model of Equation (1). Right panel: fit to experimental data according to the fluctuation-assisted tunnelling model of Equation (2).

Considering the real geometry of practical samples, physical models must be modified to take into account in particular the probability and nature of a contact between dispersoids. For example both computations and experiments show that the aspect ratio of the dispersoid plays an important role for the resulting composite properties: in particular, high aspect ratio materials (Balberg, 1987) result in a lower percolation threshold with respect to spherical particle ones (Chiolerio et al., 2010). An interesting simplification of the fluctuation-induced tunnelling for constant temperature (Kilbride et al., 2002), assuming that the CNT content is homogeneously dispersed, results in a scaling of the conductivity according to Equation 2:

$\ln \sigma = -Ap-1/3$

This means that, under constant temperature, the electrical behavior is described by fluctuations depending on the mean distance between CNTs, which determines the potential barrier height. A fit to experimental data according to Equation (2) is presented in Figure 13, right panel. The model provides a very good and universal framework to predict data for every sample considered in this chapter. Parameters extrapolated from the equation are presented in Table 3. In particular, it is possible to see from the R^2 values that the fit according to this model is more accurate. It appears that PDMS has the lowest effect on affinity, the lowest the probability that the matrix affects the dispersion geometry. When the affinity, on the contrary, is not so high, CNTs may be detached from the matrix and increasing their concentration may have a smoother effect.

Table 3.

Matrix	p_c [v.-%]	T	R^2	A	R^2
E	0.2±0.1	2.0±0.2	0.97	11±1	0.96
H1	2.0±0.1	2±1	0.90	31±2	0.99
PDMS	0.7±0.2	2.7±0.5	0.91	9±1	0.93
PVB	2.0±0.1	2±1	0.84	17±2	0.97

Properties of the PMNCs as extrapolated from models in Equation (1) and (2).

CONCLUSIONS

In conclusion we prepared four sets of NCs, using four commercial polymeric matrices (E, H1, PDMS and PVB), by dispersing an increasing amount of functionalized MWCNTs. Aim of our work was the easy preparation of cost-effective materials for electrical applications. A detailed characterization, made making use of sophisticated FEM simulations and a careful realization of a measurement setup, allowed to collect confident estimates for the resistivities and to apply physical models such as the percolation theory and the fluctuation-mediated tunneling theory. Parameters extracted from the model fitting allowed us to conclude that the lowest percolation threshold may be found for resin E; the PDMS resin shows the lowest variation of electrical properties by addition of CNTs, even though the matrix conductivity is quite high even with a very low amount of CNTs. On the contrary, resin H1 shows the higher variation of electrical resistance by addition of CNTs. Finally, PVB NCs show an interesting threshold-free behavior, consisting in a linear variation of logarithmic conductivity versus CNTs volume fraction.

REFERENCES

The Authors would like to acknowledge Dr. Salvatore Guastella and Mr. Edoardo Benassi for their precious help.

1. P. M. Ajayan, O. Stephan, C. Colliex, D. Trauth, 1994 Aligned carbon nanotube arrays formed by cutting a polymer resin-nanotube composite, Science 265 (1994), 1212 1214
2. I. Balberg, 1987 Tunneling and nonuniversal conductivity in composite materials. Physical Review Letters, 59 12 (1987), 1305 1308
3. O. Breuer, U. Sundararaj, 2004 Big returns from small fibers: a review of polymer/carbon nanotube composites, Polymer Composite 25 (2004), 630 645
4. M. Castellino, M. Tortello, S. Bianco, S. Musso, M. Giorcelli, M. Pavese, R. S. Gonnelli, A. Tagliaferro, 2010 Thermal and Electronic Properties of Macroscopic Multi-Walled Carbon Nanotubes Blocks, Journal of Nanoscience and Nanotechnology, 10 2010), 3828 3833 , 1533-4880
5. A. Chiolerio, P. Allia, A. Chiodoni, F. Pirri, F. Celegato, M. Coisson, 2007 Thermally evaporated Cu-Co top spin valve with random exchange bias. Journal of Applied Physics, 101 (2007), 123915 123911 -123915-6
6. A. Chiolerio, S. Musso, M. Sangermano, M. Giorcelli, S. Bianco, M. Coisson, A. Priola, P. Allia, A. Tagliaferro, 2010 Preparation of polymer-based composite with magnetic anisotropy by oriented carbon nanotube dispersion. Diamond and Related Materials, 17 (2008), 1590 1595
7. A. Chiolerio, L. Vescovo, M. Sangermano, 2010 Conductive UV-cured acrylic inks for resistor fabrication: models for their electrical properties.

Scientific Analysis of Nano Materials and Carbon Nanotubes 283

Macromolecular Chemistry and Physics, 211 (2010), 2008 2016

8. J. N. . Coleman, S. Curran, A. B. Dalton, A. P. Davey, B. Mc Carthy, W. Blau, R. C. Barklie, 1998 Percolation-dominated conductivity in a conjugated-polymer-carbon-nanotube composite, Physical Review B, 58 1998), R7492 R7495

9. J. N. Coleman, U. Khan, W. J. Blau, Y. K. Gun'ko, 2006 Small but strong: a review of the mechanical properties of carbon nanotube polymer composites, Carbon 44 (2006), 1624 1652

10. P. G. Collins, P. Avouris, American. Scientific, 283 283 (2000), 62

11. Gao Guan-hui, Chen Shou-gang, Xue Rui-ting, Yin Yan-sheng 2009 Study on the surface modification and dispersion of multi-walled carbon nanotubes, Advanced Material Research, Vols. 79-82, (2009), 609 612

12. A. S. Grove, 1993 Physics and Technology of Semiconductor Devices, Wiley & Sons, 1993

13. J. Hone, C. Piskoti, A. Zettl, Thermal conductivity of single-walled carbon nanotubes. Physics Review B, 59 4), 1999 R2514-6

14. N. . Hu, Karube. Yoshifumi, Yan. Cheng, Masuda. Zen, Fukunaga. Hisao, 2008 Tunneling effect in a polymer/carbon nanotube nano composite strain sensor, Acta Materialia, 56 (2008), 2929 2936

15. S. Iijima, 1991 Helical microtubules of graphitic carbon, Nature, 354 56 58

16. B. E. Kilbride, J. N. Coleman, P. Fournet, M. Cadek, A. Drury, S. Hutzler, S. Roth, W. J. Blau, 2002 Experimental observation of scaling laws for alternating current and direct current conductivity in polymer-carbon nanotube composite thin films, Journal of Applied Physics, 92 (2002), 4024 4030

17. C. H. Liu, S. S. Fan, 2007 Nonlinear electrical conducting behaviour of carbon nanotube networks in silicone elastomer. Applied Physics Letters, 90 (2007), 041905 041901 -041905-3

18. Diarmid. A. G. Mac, 2002 Synthetic metals: a novel role for organic polymers. Synth. Metals, 125 11 22 , 0379-6779

19. M. Moniruzzaman, K. I. Winey, 2006 Polymer nanocomposites containing carbon nanotubes, Macromolecules 39 (2006), 5194 5205

20. D. K. Schroder, 1990 Semiconductor Material and Device Characterization, Wiley Inter-Science Publications, 1990

21. P. Sheng, 1980 Fluctuation-induced tunnelling conduction in disordered materials. Physical Review B, 21 6 (1980), 2180 2195

22. Smits F.M., "Measurement of sheet resistivities with the four-point probe", Bell Syst. Tech. J., 1958 711 718

23. D. Stauffer, A. Aharony, 1994 Introduction to percolation theory. Taylor and Francis, London, 1994
24. M. Yu, F. , B. S. Files, S. Arepalli, R. S. Ruoff, Loading. Tensile, Ropes. of, Single. of, Carbon. Wall, Nanotubes, Mechanical. their, Physical. Properties, Letters. Review, 84 84 (2000),5552
25. http://www.nanocyl.com/en/Products-Solutions/Products/Research-Grades/Thin-Multi-Wall-Carbon-Nanotubes
26. http://www.leuna-harze.de/eindex.html
27. http://www.leuna-harze.de/eindex.html
28. http://www.dowcorning.com/applications/search/products/Details.aspx?prod=01064291&type=PROD
29. http://www3.dowcorning.com/DataFiles/090007b2815a650e.pdf
30. http://www3.dowcorning.com/DataFiles/090007b28147aa39.pdf
31. http://www.sigmaaldrich.com/catalog/DisplayMSDSContent.do

Citations

CHAPTER 1

Freeman DD, Choi K, Yu C (2012) N-Type Thermoelectric Performance of Functionalized Carbon Nanotube-Filled Polymer Composites. PLoS ONE 7(11): e47822. doi:10.1371/journal.pone.0047822.

CHAPTER 2

Choi K, Yu C (2012) Highly Doped Carbon Nanotubes with Gold Nanoparticles and Their Influence on Electrical Conductivity and Thermopower of Nanocomposites. PLoS ONE 7(9): e44977. doi:10.1371/journal.pone.0044977

CHAPTER 3

Tang ACL, Hwang G-L, Tsai S-J, Chang M-Y, Tang ZCW, et al. (2012) Biosafety of Non-Surface Modified Carbon Nanocapsules as a Potential Alternative to Carbon Nanotubes for Drug Delivery Purposes. PLoS ONE 7(3): e32893. doi:10.1371/journal.pone.0032893

CHAPTER 4

Cheng Q, Blais M-O, Harris G, Jabbarzadeh E (2013) PLGA-Carbon Nanotube Conjugates for Intercellular Delivery of Caspase-3 into Osteosarcoma Cells. PLoS ONE 8(12): e81947. doi:10.1371/journal.pone.0081947.

CHAPTER 5

McDevitt MR, Chattopadhyay D, Jaggi JS, Finn RD, Zanzonico PB, et al. (2007) PET Imaging of Soluble Yttrium-86-Labeled Carbon Nanotubes in Mice. PLoS ONE 2(9): e907. doi:10.1371/journal.pone.0000907.

CHAPTER 6

Yoon I, Hamaguchi K, Borzenets IV, Finkelstein G, Mooney R, et al. (2013) Intracellular Neural Recording with Pure Carbon Nanotube Probes. PLoS ONE 8(6): e65715. doi:10.1371/journal.pone.0065715

CHAPTER 7

Cheryl Wong Po Foo, Siddharth V. Patwardhan, David J. Belton, Brandon Kitchel, Daphne Anastasiades, Jia Huang, Rajesh R. Naik, Carole C. Perry, and David L. Kaplan Novel nanocomposites from spider silk–silica fusion (chimeric) proteins PNAS 2006 103 (25) 9428-9433; published ahead of print June 12, 2006, doi:10.1073/pnas.0601096103

CHAPTER 8

E. El-Kady, T. Mahmoud and A. Ali, "On the Electrical and Thermal Conductivities of Cast A356/Al_2O_3 Metal Matrix Nanocomposites,» *Materials Sciences and Applications*, Vol. 2 No. 9, 2011, pp. 1180-1187. doi: 10.4236/msa.2011.29159.

CHAPTER 9

Hong J, Stefanescu E, Liang P, Joshi N, Xue S, et al. (2012) Carbon Nanotubes Based 3-D Matrix for Enabling Three-Dimensional Nano-Magneto-Electronics. PLoS ONE 7(7): e40554. doi:10.1371/journal.pone.0040554

CHAPTER 10

Klinger C, Patel Y, Postma HWC (2012) Carbon Nanotube Solar Cells. PLoS ONE 7(5): e37806. doi:10.1371/journal.pone.0037806

CHAPTER 11

Liu B, Campo EM, Bossing T (2014) Drosophila Embryos as Model to Assess Cellular and Developmental Toxicity of Multi-Walled Carbon Nanotubes (MWCNT) in Living Organisms. PLoS ONE 9(2): e88681. doi:10.1371/journal.pone.0088681

CHAPTER 12

S. Samal, "Role of Temperature and Carbon Nanotube Reinforcement on Epoxy based Nanocomposites,"Journal of Minerals and Materials Characterization and Engineering, Vol. 8 No. 1, 2009, pp. 25-36.

CHAPTER 13

Wenjie Wang and N. Sanjeeva Murthy (2011). Characterization of Nano-tube- Reinforced Polymer Composites, Carbon Nanotubes - Polymer Nanocomposites, Dr. Siva Yellampalli (Ed.), ISBN: 978-953-307-498-6, In-Tech, DOI: 10.5772/20267. Available from: http://www.intechopen.com/books/carbon-nanotubes-polymer-nanocomposites/characterization-of-nanotube-reinforced-polymer-composites

CHAPTER 14

Alessandro Chiolerio, Micaela Castellino, Pravin Jagdale, Mauro Giorcelli, Stefano Bianco and Alberto Tagliaferro (2011). Electrical Properties of CNT-Based Polymeric Matrix Nanocomposites, Carbon Nanotubes - Polymer Nanocomposites, Dr. Siva Yellampalli (Ed.), ISBN: 978-953-307-498-6, InTech, DOI: 10.5772/18900. Available from: http://www.intechopen.com/books/carbon-nanotubes-polymer-nanocomposites/electrical-properties-of-cnt-based-polymeric-matrix-nanocomposites

INDEX

F

G

H

I

L

M

N

O

P